Machine Design II

A Concise Approach

As Per Latest Syllabus of UPTU, UTU, GGSIP, MDU and other Indian Universities

Machine Design II
A Concise Approach

As Per Latest Syllabus of UPTU, UTU, GGSIP, MDU and other Indian Universities

Nitin Jauhari BTech(Mech), ME(Design)

Associate Professor
Department of Mechanical Engineering
Inderprastha Engineering College
Ghaziabad, UP

CBS

CBS Publishers & Distributors Pvt Ltd

New Delhi • Bengaluru • Chennai • Kochi • Kolkata • Mumbai • Pune
Hyderabad • Nagpur • Patna • Vijayawada

Machine Design II
A Concise Approach

As Per Latest Syllabus of UPTU, UTU, GGSIP, MDU and other Indian Universities

ISBN: 978-81-239-2876-0

Published by Satish Kumar Jain and produced by Varun Jain for

CBS Publishers & Distributors Pvt Ltd

4819/XI Prahlad Street, 24 Ansari Road, Daryaganj, New Delhi 110 002, India.
Ph: 23289259, 23266861, 23266867   Website: www.cbspd.com
Fax: 011-23243014   e-mail: delhi@cbspd.com; cbspubs@airtelmail.in.
Corporate Office: 204 FIE, Industrial Area, Patparganj, Delhi 110 092
Ph: 4934 4934   Fax: 4934 4935   e-mail: publishing@cbspd.com; publicity@cbspd.com

Branches

- **Bengaluru:** Seema House 2975, 17th Cross, K.R. Road,
  Banasankari 2nd Stage, Bengaluru 560 070, Karnataka
  Ph: +91-80-26771678/79   Fax: +91-80-26771680   e-mail: bangalore@cbspd.com
- **Chennai:** 7, Subbaraya Street, Shenoy Nagar, Chennai 600 030, Tamil Nadu
  Ph: +91-44-26260666, 26208620   Fax: +91-44-42032115   e-mail: chennai@cbspd.com
- **Kochi:** Ashana House, No. 39/1904, AM Thomas Road, Valanjambalam, Eranakulam 682 018,
  Kochi Kerala
  Ph: +91-484-4059061-65   Fax: +91-484-4059065   e-mail: kochi@cbspd.com
- **Kolkata:** 6/B, Ground Floor, Rameswar Shaw Road, Kolkata-700 014, West Bengal
  Ph: +91-33-22891126, 22891127, 22891128 e-mail: kolkata@cbspd.com
- **Mumbai:** 83-C, Dr E Moses Road, Worli, Mumbai-400018, Maharashtra
  Ph: +91-22-24902340/41   Fax: +91-22-24902342   e-mail: mumbai@cbspd.com
- **Pune:** Bhuruk Prestige, Sr. No. 52/12/2+1+3/2 Narhe, Haveli
  (Near Katraj-Dehu Road Bypass), Pune 411 041, Maharashtra
  Ph: +91-20-64704058, 64704059, 32392277 Fax: +91-20-24300160   e-mail: pune@cbspd.com

Representatives

- **Hyderabad** 0-9885175004   • **Nagpur** 0-9021734563
- **Patna**   0-9334159340   • **Vijayawada** 0-9000660880

Printed at: Swastik Packagings, 506 F.I.E. Patparganj, Delhi - 92

Preface

This book deals with the design and analysis of mechanical components like gears, bearings, IC engine parts, etc. in a simple and lucid language based on actual classroom interactions. Course contents have been planned in such a way that the book (in SI Units) covers the complete course of subject for 6th semester (third year) of UPTU (Lucknow), UTU (Dehradun), GGSIP University (Delhi) and MDU (Rohtak) and other Indian universities.

The book contains the following six chapters:

Unit I: Spur gears
Unit II: Helical gears
Unit III: Worm gears
Unit IV: Sliding contact bearings
Unit V: Rolling contact bearings
Unit VI: IC engine parts

The book has been written in a simple and easy way to follow language, keeping in mind a classroom with both an average student as well as meritorious one, who can grasp the subject basics easily by self-study. Each chapter gives to the point design data tables, figures, and concepts along with numericals including problems from previous years' university examinations. A revision checklist is given at the beginning of every chapter and a brief summary of contents is given at the end so as to have a thorough understanding as well as a quick revision of the subject. Solved university papers along with unsolved problems are included in every chapters for practice.

I express my appreciation and gratefulness to CBS Publishers & Distributors, for their continued cooperation and untiring efforts in bringing out a quality book within a short time.

Though care has been taken in checking manuscript and proofreading, I shall be grateful to the readers for highlighting mistakes (if any) that might have crept in, along with the suggestions for improvement, which will be welcome and usefully incorporated in the next printing/edition of the book.

Nitin Jauhari

Acknowledgments

First and foremost, I wish to thank my parents Dr Umesh Jauhari and Mrs Chitra Jauhari for their motivation and guidance.

I feel gratitude towards my research guides Dr RK Mishra and Dr Harischandra Thakur at GBU, Greater Noida, for their continued guidance and support.

My workplace Inderprastha Engineering College, Ghaziabad, and Prof D Ganguly, Head, Department of Mechanical Engineering, deserve special mention along with my undergraduate students over the years for providing an environment conducive to good quality research and academics. I would also like to acknowledge SHIGLEY/ BHANDARI:*Design of Mechanical Equipments* and SINGH/KUMAR:*Machine Design* for reference work.

Family members including my wife Nikita and dearest daughters, Dhruvika and Navika deserve credit for their unconditional support and patience without which the work in its present form would not have been possible.

Nitin Jauhari

Contents

4. Sliding Contact Bearings

5. Rolling Contact Bearings

1

Spur Gears

REVISION CHECKLIST

Essential Points

- Gear drives and classification
- Law of gearing and gear terminology
- Gear tooth forms
- Systems of gear tooth
- Contact ratio
- Standard proportion of gear systems
- Interference and backlash
- AGMA and Indian standards
- Selection of gear materials
- Gear manufacturing methods
- Design considerations: Beam strength and wear strength
- Failure of gears
- Design of spur gears

1.1 INTRODUCTION

Gears are mechanical drives used to transmit mechanical power over a certain distance between shafts which usually involve change in speed and torque.

Mechanical drives are required placement between prime mover (motor) and the operating machine. Gears usually transmit power by means of engagement or meshing rather than by friction, which differentiates it from mechanical drives like belt drive and rope drive. Gear classification on the basis of tooth profile classifies gears as: Spur, helical, bevel and worm gears.

This Chapter deals with a precise understanding of various tooth forms in spur gears, system of gear teeth and various terminologies of gears along with design considerations of spur gears involving beam strength, dynamic tooth load, wear strength, etc. and finally the failure criteria and AGMA and Indian standards of specifications for these types of gears.

For a quick review, a gear drive with straight or parallel tooth, or involving teeth cut parallel to the axis of the shaft is classified as spur gear. The profile of gear tooth is in involute curve shape and remains identical along the entire width.

Spur gears are used for power transmission between parallel shafts and impose radial loads (acting towards center) on the shafts.

1.2 BASIC CONCEPTS

Gear Drives

Gears are toothed wheels which transmit power and motion from one shaft to another by means of successive engagement of teeth. Gear drives offer following advantages.

- Center distance between shafts is relatively small, which results in compact construction.
- Can transmit large power.
- Velocity ratio remains constant.
- Can transmit motion at very low velocity, which is not possible with the belt drives.
- Efficiency of gear drives is very high.
- Gear shifting allows changing the velocity ratio over a wide range.

Disadvantages include high cost and maintenance costs. Manufacturing is specialized and is quite complicated. Requires precise shaft alignment and proper lubrication and cleanliness.

Gear Classification

Gears are classified into 4 groups:

1. Spur
2. Helical
3. Bevel, and
4. Worm gears

Spur gears consist of teeth parallel to shaft axis and shafts need to be parallel for power transmission. Gear tooth profile is 'involute' (see Section 1.1.1) and remains identical along the entire width of the gear wheel.

The loads transmitted on shafts using spur gears are "radial loads".

Helical gears consist of teeth cut at an angle with the axis of the shaft. Tooth form is 'involute' (see Section 1.1.1) similar to spur gears. The only difference being an involute profile is lying on a plane perpendicular to tooth element.

Magnitude of helix angle remains same for pinion (driving gear) and driven gear, but sense of helix is opposite with a right hand pinion meshing with a left hand gear. These gears transmit radial and thrust load on the shafts.

A kind/type of helical gear consisting of opposite hands of helix in one gear only, is called Herringbone gear, which results in equal and opposite thrust reactions balancing each other and no thrust is loaded on the shaft. Like helical gears, Herringbone gears are too used only for parallel shaft.

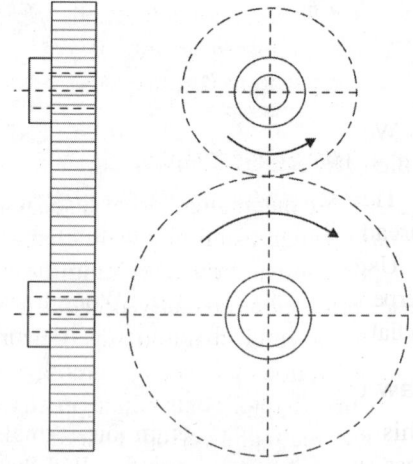

Fig. 1.1 Spur gears

Bevel gears are used for power transmission between shafts which are at right angles to each other. Angle can be slightly flexible too. Tooth of bevel gears can be cut straight or spiral.

These gears too like helical gears impose radial and thrust loads on shafts. Bevel gears have the shape of a truncated cone.

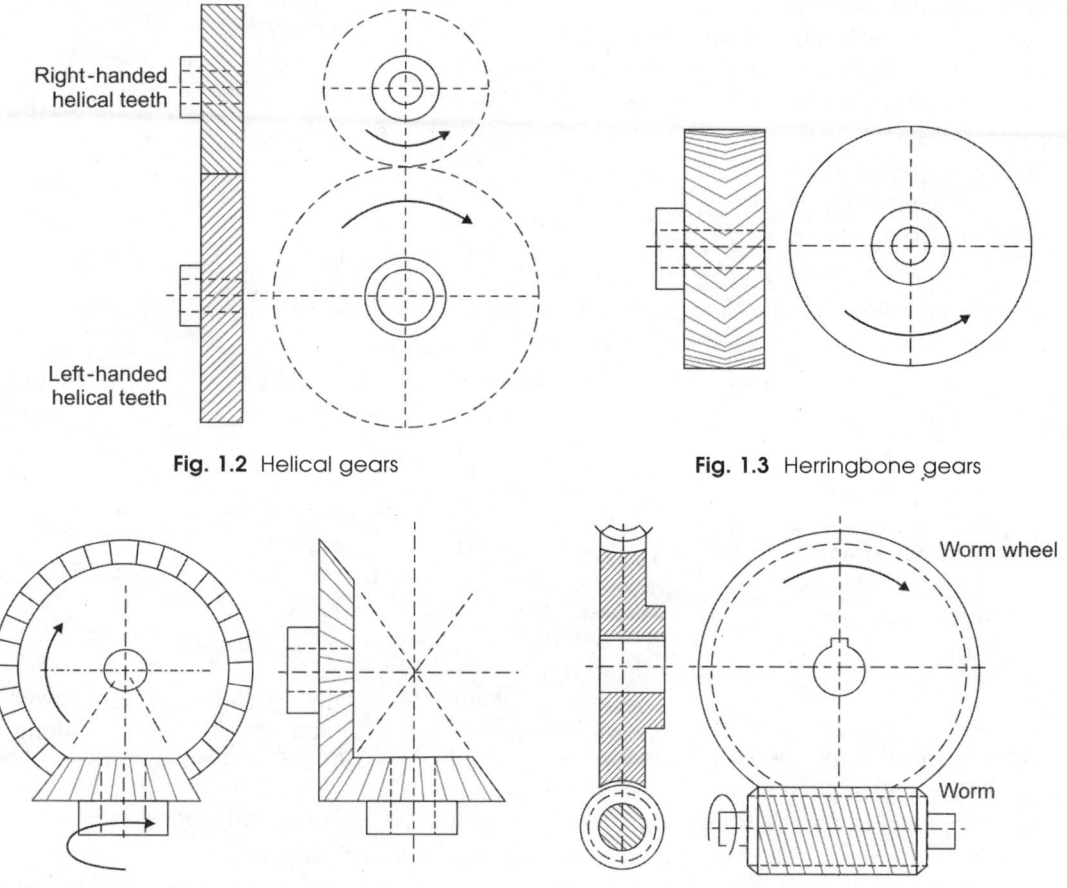

Right-handed helical teeth

Left-handed helical teeth

Fig. 1.2 Helical gears

Fig. 1.3 Herringbone gears

Fig. 1.4 Bevel gears

Fig. 1.5 Worm gears

Worm wheel

Worm

Worm gears consist of a worm and a worm wheel. Worm being in form of a threaded screw, which meshes with the matching wheel.

Threads on worm usually have a small lead (linear) distance moved in one revolution of thread and can be single- or multi-start threads.

Usually used for power transmission between shafts, whose axes do not intersect and are perpendicular to each other. Worm transmits high thrust load whereas worm wheel transmits high radial loads on the shafts. These have an advantage of providing very high speed reduction ratios.

Law of Gearing

This law states that "common normal to the tooth profile at the point of contact must always pass through a fixed point, called the pitch point, so as to provide a constant speed/velocity ratio". The profiles of the teeth of mating gears are designed so as to provide constant speed ratio. This condition is called profile having conjugate action.

This statement can also be stated as the common normal passing through the point of contact of two tooth profiles must divide the center distance into portions inversely proportional to the angular speed of the gears.

Considering a pair of involute gears following conjugate action, have teeth coming into contact at Q.

As the pinion rotates at speed ω_1, the pinion teeth push gear tooth to impart it an angular speed ω_2. Common normal passes through $Q \rightarrow MN$.

So circumferential speed at point Q, with respect to centers O_1 and O_2 of the pinion and gear respectively are:

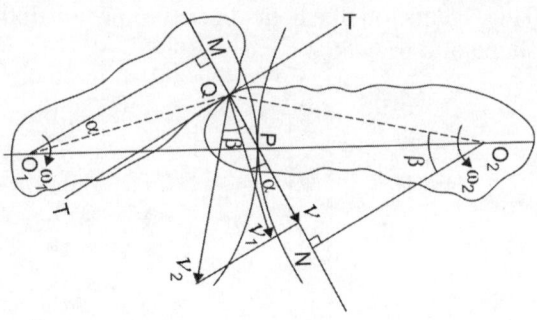

Fig. 1.6 Law of gearing

$$V_1 = (O_1Q)\omega_1 \text{ and } V_2 = (O_2Q)\omega_2$$

Condition for teeth in mesh to remain in constant contact, the velocity components V_1 and V_2 along the common normal MN must be equal, i.e.

$$V_1 \cos \alpha = V_2 \cos \beta$$

$$\Rightarrow (O_1Q) \, \omega_1 \cos \alpha = (O_2Q) \, \omega_2 \cos \beta$$

So

$$\frac{\omega_1}{\omega_2} = \frac{(O_2Q)\cos\beta}{(O_2Q)\cos\alpha} = \frac{O_2N}{O_1N}$$

From similar triangles O_1MP and O_2NP,

$$\frac{O_2N}{O_1M} = \frac{O_2P}{O_1P}$$

or

$$\frac{\omega_1}{\omega_2} = \frac{O_2P}{O_1P} = \text{Constant}$$

This proves that the common normal passes through a fixed point P on the center line O_1O_2. This point is called pitch point.

Line segment between 'M and N' on common normal is called 'line of action'.

Circles of radius O_1P and O_2P are termed as pitch circles of the gears.

(**Note:** It has been seen that only involute and cycloidal curves satisfy the fundamental law of gearing. Also pressure angle being constant throughout contact in involute profile gear tooth, makes it a better choice as compared to cycloidal profile tooth in which pressure angle keeps on varying during the contact.)

Gear Terminology

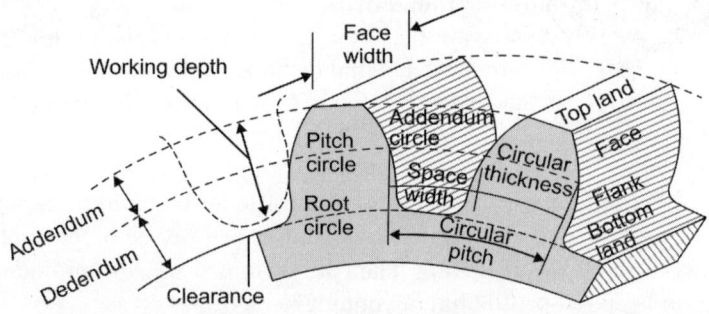

Fig. 1.7 Gear nomenclature

1. Pitch circle: The distance between corresponding sides of the adjacent teeth of a gear is measured about pitch circle. This can be considered as the circle drawn with radius as

line joining the center to the pitch point (as per the law of gearing). The circle on which the distance between corresponding sides of adjacent teeth becomes equal to rack tooth pitch is called pitch circle.

Diameter of this circle called pitch circle diameter (PCD) is *d*.

2. Module (*m*): It is the ratio of the pitch circle diameter in millimeter to the number of teeth:

$$\text{Module } (m) = d/T$$

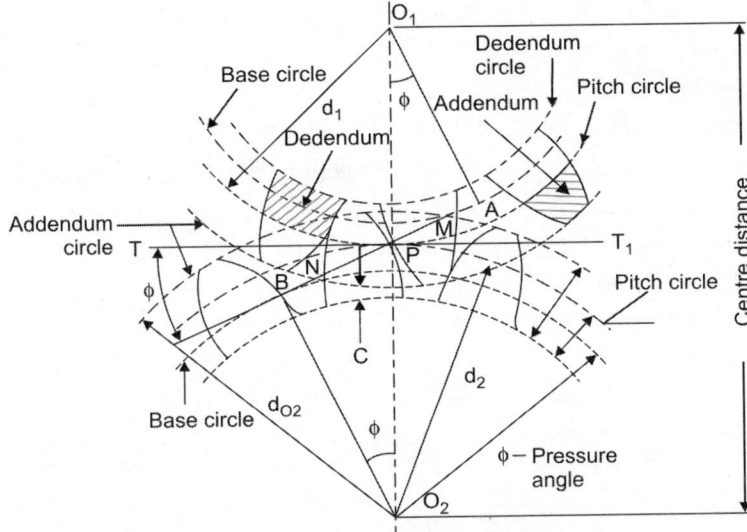

Fig. 1.8 Gear teeth in mesh

(**Important:** For two gears in mesh, for power transmission, it is essential that their modules are equal along with follow of conjugate action or law of gearing by gear teeth.)

3. Circular pitch: It is the distance between corresponding sides of two adjacent teeth of a gear measured on the pitch circle. It is expressed as

$$P_c = \frac{\pi d_1}{T_1} \text{ or } \frac{\pi d_2}{T_2}$$

D_2 and D_2 being pitch circle diameters of pinion and driven gears respectively and T_1 and T_2 being their number of teeth.

4. Diametrical pitch: Ratio of number of teeth and the pitch circle diameter:

$$P_d = \frac{T}{d} = \frac{\pi}{P_c}$$

5. Addendum and addendum circle: Circle which bounds or touches the outer ends of the teeth is addendum circle, with addendum '*a*' being the radial distance between the pitch circle and the addendum circle.

(**Note:** Its standard value is taken as $a = 1$ m, i.e. one module.)

6. Dedendum and dedendum circle: The innermost circle which bounds the bottom of the teeth is called dedendum circle or root circle.

Radial distance between the pitch circle and the dedendum circle is called '*d*'. Its standard value being 1.25 m.

7. Pressure angle (ϕ): Common normal at the point of contact to the mating involute curves is the line of action.

   The acute angle between the line of action and tangent to the pitch circle at the pitch point is called pressure angle.

   Common values of pressure angle for gears with involute teeth are $14\frac{1}{2}°$, $20°$ and $25°$.

8. Base circle: Auxiliary circle through which the involute (tooth profile) is generated or started is called base circle.

   (**Note:** Line of action is tangent to the base circles of mating gears.)

9. Clearance: The amount by which the dedendum of a gear exceeds the addendum of its mating gear.

   Clearance $(C) = d - a = 0.25$ m for standard tooth.

10. Tooth thickness: Thickness of the tooth measured along the pitch circle.

11. Tooth height: Height or the whole depth of the tooth is the sum of addendum and dedendum, i.e. tooth height $(H) = a + d$.

12. Backlash: Difference between the tooth space and the tooth thickness, as measured on pitch circle required for proper teeth meshing and no jamming is called as backlash.

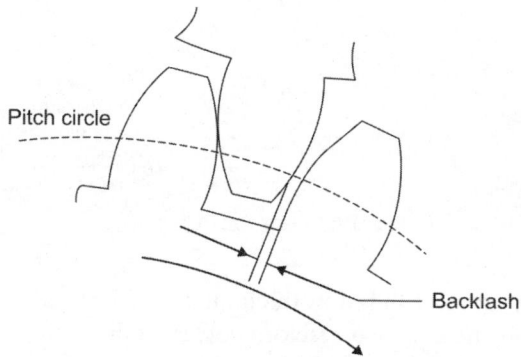

Fig. 1.9 Backlash

13. Face and flank: Portion of the tooth between pitch circle and the addendum circle (sideways) is called tooth face whereas portion of the tooth between pitch circle and the root circle (sideways) is called flank.

14. Arc of contact: Path traced by a point on the pitch circle from beginning to the end of the engagement of a given pair of teeth is called arc of contact. It can be considered consisting of two parts.

    a. Arc of approach: Portion of the path of contact from the beginning of the engagement to the pitch point.

    b. Arc of recess: Portion of the path of contact from the pitch point to the end of engagement of a pair of teeth.

1.3 TOOTH FORMS

Power transmission involving gears requires meshing or engagement of gears which require continuous engagement and disengagement of gear tooth in mesh between driving and driven gears.

This fundamentally must satisfy law of gearing (Fig. 1.1) which states that—'the common normal to the tooth profile at the point of contact always passes through a fixed point called as the pitch point, in order to obtain a constant velocity ratio'.

To satisfy law of gearing, a gear tooth curvature must be designed as 'involute' or 'cycloidal' form.

1.3.1 Involute Curve Tooth Form

Involute is a curve traced by a point on a line as the line rolls without slipping on a circle. In case of involute profile, the common normal at the point of contact always passes through the pitch point P and maintains a constant inclination α with common tangent to two pitch circles. This angle α is called pressure angle and it remains constant in involute teeth form. Involute tooth form is most widely used in gear tooth design.

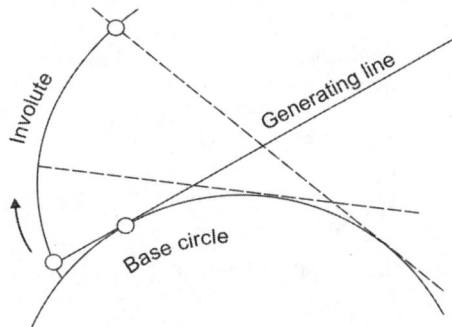

Fig. 1.10 Involute curve

1.3.2 Cycloidal Curve Tooth Form

A cycloid is a curve traced by a point on the circumference of a generating circle as it rolls without slipping along the inside (hypocycloid) and outside (epicycloid) of another circle.

Though less used as compared to involute tooth form, due to variation in pressure angle during meshing of gear teeth and thereby following of law of gearing at only specific points during engagement of teeth, cycloidal curve tooth form finds some applications like spring driven watches and clocks.

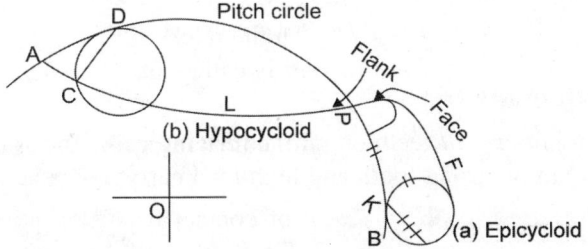

Fig. 1.11 Cycloidal curve form: (a) Epicycloidal; (b) Hypocycloidal

1.3.3 Advantages and Disadvantages

Involute tooth form maintains a constant pressure angle α during the gear teeth meshing or engagement which is not so in cycloidal tooth form. Also profile of involute teeth is made of a single curve and only one cutter is necessary to manufacture one complete set of pinion (driving gear) and gear (driven), thereby resulting in reduction of manufacturing cost which is not so in cycloidal teeth form which is difficult to manufacture. Cycloidal tooth form on the other hand, involves mating of a convex flank on one tooth coming in contact with concave flank of the other tooth. This increases the contact area and also the wear strength. This is in contrast to the involute tooth form involving contact between two convex surfaces on mating teeth, resulting in smaller contact area and lower wear strength. Other advantages of cycloidal tooth form include satisfactory operation or power transmission with very small number of teeth, along with the absence of interference (Section 1.5), and undercutting which is usually seen in involute tooth form.

1.4 SYSTEM OF GEAR TEETH

Gear teeth system with mostly involute profile, is classified as (on the basis of pressure angle)

- $14\frac{1}{2}°$ full depth involute
- $20°$ full depth involute
- $20°$ stub involute system

Involute profile is used on account of:

- Same module and pressure angle gears are interchangeable and can be machined by a single tool.
- Involute profile gear tooth follows law of gearing at any center distance and pitch point remains fixed.

1.4.1 $14\frac{1}{2}°$ Full Depth Involute

This gear system requires minimum no. of teeth on pinion (driving gear) to be 23, so as to avoid interference (Section 1.5) and undercutting (Section 1.5). Advantage of use being quietness of operations of power transmission.

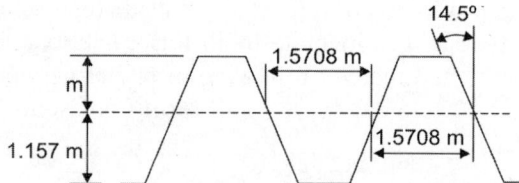

Fig. 1.12 $14\frac{1}{2}°$ full depth involute

1.4.2 $20°$ Full Depth Involute

This system requires minimum 17 teeth on pinion (driving gear) and is most widely used and recommended on account of strong tooth and high load carrying capacity.

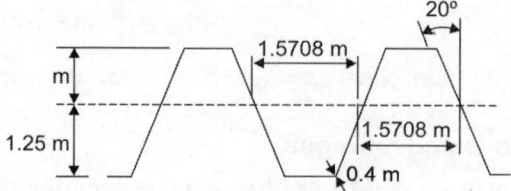

Fig. 1.13 $20°$ full depth involute

With increase of pressure angle, interference (Section 1.5) and undercutting (Section 1.5) is reduced. A significant advantage of 20° full depth involute system over $14^1/_2$° system is, teeth being more broad at root (see Fig. 1.13), increases load carrying capacity and thereby overall strength and greater line of contact. The only disadvantage being reduced duration of contact during teeth engagement.

1.4.3 20° Stub Involute

This system requires minimum 14 teeth on pinion for avoidance of interference and undercutting. It has shorter addendum (see basic concepts in Section 1(d)5) and dedendum (see basic concepts in Section 1(d)6) thereby reducing interference and undercutting slightly more and has more stronger teeth.

Disadvantages of this gear system involves more vibrations due to insufficient overlap and lesser contact ratio (Section 1.3) due to shorter addendum.

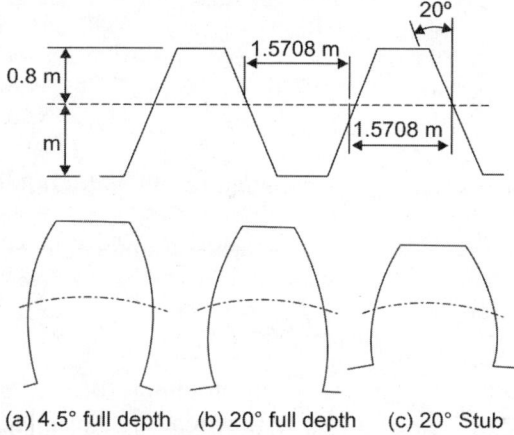

(a) 4.5° full depth   (b) 20° full depth   (c) 20° Stub

Fig. 1.14 20° stub involute system

1.5 CONTACT RATIO

The number of teeth in actual simultaneous contacts during gear mesh or engagement is called contact ratio. For smooth power transmission, usually its value is taken as 1.2, whereas for industrial gear box as 1.6–1.7. Overlapping of gear teeth is required to ensure that minimum 2 teeth are always in contact, so that contact between 1 pair of gears continues till contact remains or during gear tooth engagement.

1.6 STANDARD PROPORTIONS OF GEAR SYSTEMS

Standard proportions of gear systems (in terms of gear module m)

	$14^1/_2$° full depth involute	*20° full depth involute*	*20° stub involute*
Addendum	1 m	1 m	0.8 m
Dedendum	1.25 m	1.25 m	1 m
Working depth	2 m	2 m	1.6 m
Minimum total depth	2.25 m	2.25 m	1.8 m
Tooth thickness	1.5708 m	1.5708 m	1.5708 m
Minimum clearance	0.25 m	0.25 m	0.2 m
Fillet radius at root	0.40 m	0.40 m	0.40 m

1.7 AGMA AND INDIAN STANDARDS

Table 1.1 Proportions of standard involute teeth (in terms of module m)

	14.5° full depth system	*20° full depth system*	*20° stub system*
Pressure angle	14.5°	20°	20°
Addendum	m	m	0.8 m
Dedendum	1.157 m	1.25 m	m
Clearance	0.157 m	0.25 m	0.2 m
Working depth	2 m	2 m	1.6 m
Whole depth	2.157 m	2.25 m	1.8 m
Tooth thickness	1.5708 m	1.5708 m	1.5708 m

Table 1.2 Recommended series of module (min)

Choice 1 (preferred)	1.0	1.25	1.5	2.0	2.5	3.0	4.0
	5.0	6.0	8.0	10	12	16	20
Choice 2	1.125	1.375	1.75	2.25	2.75	3.5	4.5
	5.5	7	9	11	14	18	

Table 1.3 Number of teeth (z_{min}, minimum) on pinion for different tooth systems

Pressure angle (α)	14.5°	20°	25°
z_{min} (theoretical)	32	17	11
z_{min} (practical)	27	14	9

Table 1.4 Values of the Lewis form factor Y for 20 full depth involute

z	Y	z	Y	z	Y
15	0.289	27	0.348	55	0.415
16	0.295	28	0.352	60	0.421
17	0.302	29	0.355	65	0.425
18	0.308	30	0.358	70	0.429
19	0.314	32	0.364	75	0.433
20	0.320	33	0.367	80	0.436
21	0.326	35	0.373	90	0.442
22	0.330	37	0.380	100	0.446
23	0.333	39	0.386	150	0.458
24	0.337	40	0.389	200	0.463
25	0.340	45	0.399	300	0.471
26	0.344	50	0.408	Rack	0.484

Table 1.5 Service factor for speed reduction gearboxes

Working characteristics (Driving machine)	*Working characteristics of driven machine*		
	Uniform	*Moderate shock*	*Heavy shock*
Uniform	1.00	1.25	1.75
Light shock	1.25	1.50	2.00
Medium shock	1.5	1.75	2.25

Table 1.6 Examples of driving machine with different working characteristics

Character of operation	Driving machines
Uniform	Electric motor, steam turbine, gas turbine
Light shock	Multi-cylinder internal combustion engine
Medium shock	Single cylinder internal combustion engine

Table 1.7 Recommended number of arms for long gears

Pitch diameter (mm)	Number of arms (n)
300–500	4
500–1,500	6
1,500–2,400	8
>2,400	10–12

Table 1.8 Some more examples of driven machines with different working characteristics

Character of operation	Driven machines
Uniform	Generator, belt conveyor, platform conveyor, light elevator, electric hoist, feed gears of machine tools, ventilators, turbo-blower, mixer for constant density material.
Medium shock	Main drive to machine tool, heavy elevator, turning gear of crane, mine ventilator, mixer for variable density material, multi-cylinder piston pump, feed pump.
Heavy shock	Press, shear, rubber mill, rolling mill drive, power shovel, heavy centrifuge, heavy feed pump, rotary drilling apparatus, briquette press, pug mill.

Table 1.9 Values of deformation constant C (N/mm^2)

Materials Gear material	14.5° full depth teeth	20° full depth teeth	20° stub teeth
Grey C.I.	5,500	5,700	5,900
Grey C.I.	7,600	7,900	8,100
Steel	11,000	11,400	11,900

Table 1.10 Tolerances on the adjacent pitch

Grade	e (micra)
1	0.80 + 0.06 ϕ
2	1.25 + 0.10 ϕ
3	2.00 + 0.16 ϕ
4	3.20 + 0.25 ϕ
5	5.00 + 0.40 ϕ
6	8.00 + 0.63 ϕ
7	11.00 + 0.90 ϕ
8	16.00 + 1.25 ϕ
9	22.00 + 1.80 ϕ
10	32.00 + 2.50 ϕ
11	45.00 + 3.55 ϕ
12	63.00 + 5.00 ϕ

Table 1.11 Values of bending stress factor S_b

Material	S_b
Phosphor-bronze (centrifugal/cast)	7.00
Phosphor-bronze (sand-cast and chilled)	6.40
Phosphor-bronze (sand-cast)	5.00
0.4% carbon steel-normalized (40C8)	14.10
0.55% carbon steel-normalized (55C8)	17.60
Case-hardened carbon steels (10C4, 14C6)	28.20
Case-hardened alloy steels (16Ni80Cr60 and 20Ni2Mo25)	33.11
Nickel-chromium steels (13Ni3Cr80 and 15Ni4Cr1)	35.22

Table 1.12 Values of modulus of elasticity and Poisson's ratio for gear materials

Material	Modulus of elasticity (N/mm²)	Poisson's ratio
Steel	206,000	0.3
Cast steel	202,000	0.3
Spheroidal cast iron	173,000	0.3
Cast tin bronze	103,000	0.3
Tin bronze	113,000	0.3
Grey cast iron	118,000	0.3

WORKED EXAMPLES

1. A pair of spur gears consists of a 20 teeth pinion meshing with a 120 teeth gear. The module is 4 mm.

Calculate:

  i. The center distance
  ii. The pitch circle diameters of the pinion and the gear
  iii. The addendum and dedendum
  iv. Teeth thickness
  v. Clearance
  vi. Gear ratio

SOLUTION:

As addendum $(a) = \dfrac{m\left(T_p + T_g\right)}{2}$

Given: $T_p = 20$, $T_g = 120$, $m = 4$ mm

So $a = \dfrac{4(20 + 120)}{2} = 280$ mm

So d_p = Pinion pitch circle diameter = $m \times T_p = 4 \times 20 = 80$ mm

$d_g = m\,T_g$ = Gear (driven) pitch circle diameter = $m\,T_g = 4\,(120) = 480$ mm

dedendum = $1.25m = 1.25 \times 4 = 5$ mm

teeth thickness = $1.5708m = 1.5708 \times 4$
$\qquad\qquad\qquad = 6.2832$ mm

Clearance = $0.25m = 0.25 \times 4 = 1$ mm

Gear ratio $= \dfrac{T_g}{T_p} = \dfrac{120}{20} = 6$

2. A pinion with 25 teeth and rotating at 1,200 rpm, drives a gear which rotates at 200 rpm. The module is 4 mm. Calculate the center distance between the gears.

SOLUTION:

Speed ratio $= \dfrac{1,200}{200} = 6$

Also speed ratio $= \dfrac{T_g}{T_p} = 6$

$\Rightarrow \qquad T_g = 6 \times 25 \qquad\qquad [\because T_p = 25]$

$\Rightarrow \qquad T_g = 150$

module $(m) = 4$ mm

also $\qquad m = \dfrac{d_p}{T_p} = \dfrac{d_g}{T_g}.$

$\Rightarrow \qquad d_p = m \times T_p = 4 \times 25 = 100$ mm

$\qquad\qquad d_g = m \times T_g = 4 \times 150 = 600$ mm

So center distance between gears $= \dfrac{d_p + d_g}{2} = \dfrac{100 + 600}{2} = 350$ mm

3. A pair of spur gears with a center distance of 495 mm is used for a speed reduction of 4.5:1. The module is 6 mm. Calculate the number of teeth on the pinion and the gear.

SOLUTION:

Center distance $= \dfrac{1}{2}\left(d_p + d_g\right) = 495$ mm $\qquad\qquad$ ...(i)

Speed reduction $= \dfrac{N_p}{N_g} = \dfrac{T_g}{T_p} = 4.5:1 \qquad\qquad$ ...(ii)

[Subscript $p \Rightarrow$ pinion (driving gear), $g \Rightarrow$ gear (driven)]

$\qquad\qquad$ Module $(m) = \dfrac{d_p}{T_p} = \dfrac{d_g}{T_g} = 6 \qquad\qquad$ ...(iii)

$\qquad\qquad$ To calculate $\rightarrow T_p$ and T_g

from (i) $\rightarrow d_p + d_g = 990$

from (iii) $\rightarrow d_p = 6T_p, d_g = 6T_g$

Substituting this in (i) $\Rightarrow 6T_p + 6T_g = 990$

$\Rightarrow \qquad\qquad T_p + T_g = 165 \qquad\qquad$ ...(iv)

from (ii) $\qquad \dfrac{T_g}{T_p} = 4.5 \Rightarrow T_g = 4.5\, T_p$

Substituting this in (iv) $\Rightarrow T_p + 4.5\, T_p = 165$

or
$$5.5\, T_p = 165 \Rightarrow T_p = \frac{165}{5.5} = \frac{33}{1.1} = 30$$

$\Rightarrow$ $\quad T_g = 4.5 \times 30 = 135.0$

So $\quad T_p = 30,\ T_g = 135$

4. The pitch circles of a train of spur gears are shown in figure. Gear A receives 3.5 kW power at 700 rpm through its shaft and rotates in the clockwise direction. Gear B is the idler gear while gear C is the driven gear. The number of teeth on gears A, B and C are 30, 60 and 40 respectively, while the module is 5 mm. Calculate:

   i. The torque on each gear shaft

  ii. The components of gear tooth forces.

Assume 20° involute system for the gears.

SOLUTION:

$$m = \frac{d_A}{T_A} = \frac{d_B}{T_B} = \frac{d_C}{T_C}$$

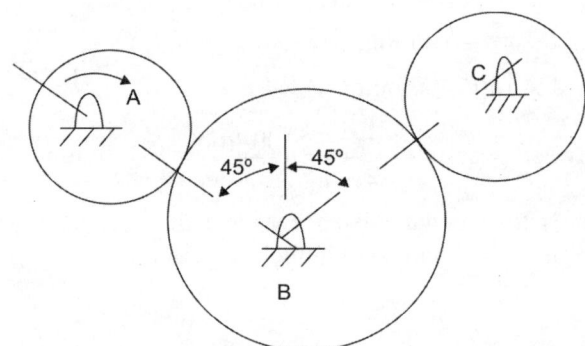

$\Rightarrow d_A = 5 \times 30 = 150$ mm, $d_B = 5 \times 60 = 300$ mm, $d_C = 5 \times 40 = 200$ mm

Torque on gear shaft $A = (M_t)_A = \dfrac{60 \times 10^6 \times P_{kW}}{2\pi n_A} = \dfrac{60 \times 10^6 \times 35}{2\pi \times 700}$

$\Rightarrow (M_t)_A = 47{,}746.48$ N-mm

Gear B being idler gear, so does not transmit any torque to its shaft. Therefore,
$$(M_t)_B = 0.$$

Since the same power is transmitted from gear A to gear C.

$\Rightarrow \quad (M_t)_A \times n_A = (M_t)_C \times n_C \quad$ [n_A and $n_C \to$ Speed in rpm of gears A and C respectively]

$\Rightarrow \quad (M_t)_C = (M_t)_A \times \left(\dfrac{n_A}{n_C}\right)$

$$= 47{,}746.48 \times \frac{40}{30} = 63{,}661.98 \text{ N-mm.}$$

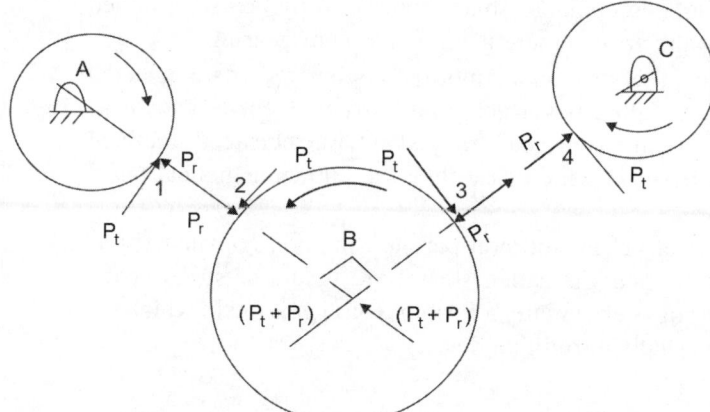

The components of the gear tooth force between gears A and B are given by,

$$(P_t)_{AB} = \frac{2(M_t)_A}{d_A} = \frac{2 \times 47,746.48}{150} = 636.62 \text{ N} \qquad \text{[as } (M_t)_A = P_t \times d/2]$$

$$(P_r)_{AB} = (P_t)_{AB} \tan \alpha = 636.62 \tan (20) = 231.71 \text{ N}$$

Since gear B is idler, whatever torque it receives from gear A is transmitted to gear C. Therefore,

$$(P_t)_{AB} = \frac{d_B}{2} = (P_t)_{BC} \times \frac{d_B}{2}$$

$\Rightarrow \qquad\qquad (P_t)_{AB} = (P_t)_{BC} = P_t$

$\Rightarrow (P_t)_{AB} = (P_t)_{AB} \Rightarrow$ since the tangential components are equal, so the radial components ($P_t \tan \alpha$) must be equal.

PREVIOUS YEAR UNIVERSITY QUESTIONS

1. Show that conjugate to an involute profile is another involute profile.

SOLUTION: Involute profile satisfies the fundamental condition of law of gearing (common normal at the point of contact between a pair of teeth must always pass through the pitch point), or the condition of law of gearing is fulfilled by teeth of involute form. Two curves of any shape that fulfill the law of gearing can be used as the profiles of teeth. In other words an arbitrary shape of one of the mating teeth can be taken and applying the law of gearing the shape of other can be determined. Such gears are said to have conjugate teeth.

Because involute profile satisfies law of gearing its conjugate will also satisfy law of gearing and conjugate profile will be the corresponding form of involute. So formed conjugate of involute will be another involute profile.

2. Give an account of gear classification. Name the gear that can be used for non-parallel non-intersecting shafts.

SOLUTION: Gears may be classified as follows:

    i. According to the position of areas of the shaft:

        The areas of two shafts between which the motion is to be transmitted, may be:

        (a) Parallel, (b) intersecting and (c) non-parallel and non-intersecting.

Two parallel and coplanar shafts connected by gears are called spur gears and arrangement is known as spur gearing.

Two non-parallel or intersecting, but coplanar shafts connected by gears are called bevel gears and arrangement is known as bevel gearing like spur gears may also have their teeth inclined to face of bevel in which case they are known as helical bevel gears.

Two non-intersecting and non-parallel, i.e. non-coplanar shafts connected by gears is called skew bevel gears or spiral gears. Worm gearing is essentially a form of spiral gearing in which the shafts are usually at right angles.

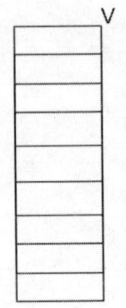

Straight spur gears

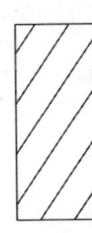

Helical gears

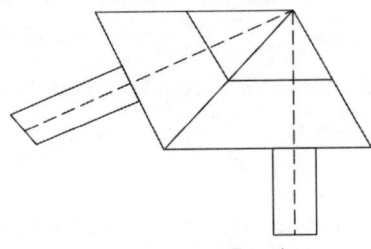

Bevel gears

ii. According to peripheral velocity of gears:

According to peripheral velocity, gears may be classified as:

   a. Low velocity gears: Velocity less than 3 m/sec.

   b. Medium velocity: Gears having velocity between 3 and 15 m/sec.

   c. High velocity gears: Gears having velocity greater than 15 m/sec.

iii. According to the type of gearing:

According to the type of gearing, gears are classified as:

   (a) External gearing, (b) internal gearing and (c) rack and pinion

iv. According to the position of teeth on gear surface:

The teeth on gear surface may be: (a) Straight, (b) inclined and (c) curved.

Non-parallel, non-intersecting shafts: Gears that can be used for non-parallel, non-intersecting shafts, i.e. non-coplanar shafts connected by gear are called skew bevel gears or spiral gears.

3. Give an account of failure of gears.

SOLUTION: See Section 1.13

4. What condition must be satisfied in order, so that a pair of spur gears may have a constant velocity ratio?

SOLUTION: In order to have a constant angular velocity ratio for all positions of the wheels, pitch point 'p' must be the fixed point for two wheels. In other words, the common normal at the point of contact between a pair of teeth must always pass through the pitch point. This is a fundamental condition which must be satisfied while designing the profiles for the teeth of gear wheels. It is also known as law of gearing.

5. What is conjugate action? How is it achieved in the case of involute gears?

SOLUTION: Any arbitrary shape of the teeth can be chosen for profile of teeth of one of the two gears in mesh and the profile for the other may be determined to satisfy the law of gearing.

Such teeth are called conjugate teeth. Theoretically such profile teeth will transmit the desired motion, but objection to such random profiles is the obvious difficulty of manufacture, standardization and cost of production. Therefore, conjugate teeth are not in normal use. Commonly used forms of teeth which satisfy the law of gearing arc:

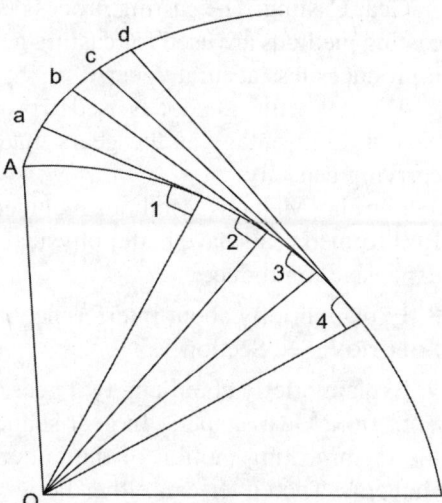

i. Involute profile teeth
ii. Cycloidal profile teeth

Construction of an involute: Let the starting point of the involute be *A*. Equal lengths *A*-1, 12, 23, 24 are marked as the base circle. Erect perpendiculars to the radii at 1, 2, 3 and 4. Lengths 1*a*, 2*b*, 3*c*, 4*d* are made equal to the corresponding are is the desired involute.

6. Enumerate advantages and disadvantages of stub gear tooth.

SOLUTION: In stub tooth system, the tooth has less working depth, usually 20% less than full depth system.

The addendum is made shorter. Stub tooth system has the following advantages:

i. Due to the shorter addendum: The interfering portion of the tooth is removed and hence there will be less interference in stub system.
ii. The tooth becomes stronger as the lever arm of the beginning moment on tooth is reduced.
iii. Machining cost is reduced as compared to full depth system.

Stub tooth system is advantageous mostly in gears with small number of teeth. With large number of teeth, the full depth system performs better. This is the only disadvantage of stub tooth system.

7. Briefly enumerate gear manufacturing methods.

SOLUTION: The various manufacturing methods used for gears can be listed under the following headings:

i. Metal removal methods
ii. Casting methods
iii. Forming methods

Metal removal methods: In these methods, the metal is removed from the gear blanks to produce the gears. Under this the main gear cutting methods are:

a. Profiling
b. Generation

In profiling method, the space between the teeth are cut by means of disc cutter or end milling cutter. Only spur, helical and straight bevel gears are being manufactured by this method.

In generating process, teeth are formed in a series of passes by a generating tool. Under this heading following methods are used:

a. Hobbing
b. Shaping
c. Broading
d. Punching
e. Shear cutting

Gear Casting: The casting process is used to make toothed gears in one operation. The sand casting methods are used for casting gear, for form machinery and hand operated machinery as it produces less accurate gears.

The die casting method is used to produce large quantities of small gears. But casting methods have a disadvantage as the gears made out of these methods are not suitable for high load carrying capacity.

Forming Methods: Methods included under this are "Cold drawing" and " Rolling forming". Roll formed gears have better physical properties than cut teeth and in large quantities, are less expensive to produce.

8. Explain briefly about interference in involute gear teeth and its remedy.

SOLUTION: See Section 1.5

9. Explain briefly about any two gear cutting methods and indicate their relative advantages.

SOLUTION: Gear shapers: In gear shapers, the cutter reciprocates rapidly. The teeth are cut by the reciprocating motion of the cutter and because of this, these machines are called "gear shapers". The cutter can either be a 'rack-type cutter', or a 'rotary pinion type cutter'. The main drawback of 'rack-type' cutter is that once the length of rack is covered by the gear blank, the cutting process is interrupted to index the blank back to starting point. In the case of 'rotary pinion type cutter', such an indexing is not required, therefore, this type is more productive and so common.

Cutting takes place either on the downward stroke or on the upward stroke of the cutter (depending upon the design of the machine) during each return stroke, or the cutter is withdrawn from the blank. Its purpose is to prevent rubbing and resulting wear of the cutting edges of the cutter and damage to the tooth profiles of the gear being cut.

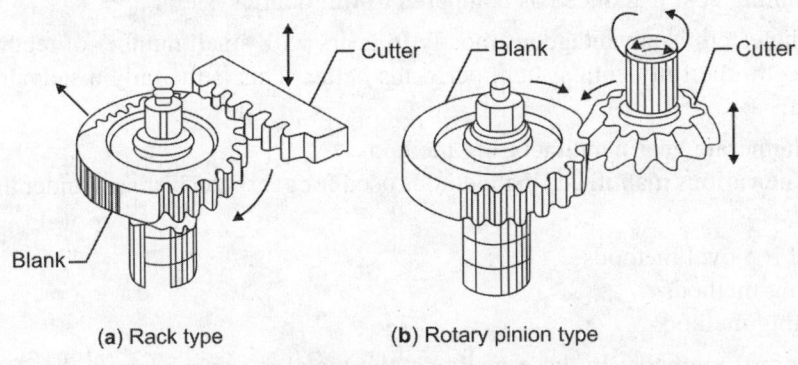

(a) Rack type   (b) Rotary pinion type

Gear shaper cutters

To start cutting a gear, the tool is fed into the gear blank before each cutting stroke. When the required depth is reached, the inward feed stops and the cutter and blank slowly start rotating as if they were in mesh. The gear is cut in one complete rotation of the blank. The rotary motions of the cutter and the blank are coordinated through change gears, so that the surface speeds of the cutter and the blank are the same. The cutter reciprocates at about 100 strokes per minute for the average job and the strokes can be up to 2,000 per minute for fine tooth gears.

Advantages

1. The cutter has a generated profile which has more accurate shape than some other cutters, so the gears produced are accurate.

2. The method is suitable for medium and large batch production.
3. Because the cutting stroke can be adjusted, gear shapers are particularly suitable for cutting cluster gears.
4. Cutter can be used for cutting all spur gears of the same module, irrespective of number of teeth on the gear.
5. The method is versatile and can cut spur, helical Herringbone, internal and cluster gears, racks, splines, etc.

Limitations

1. Separate helical guide is required for cutting helical gear of particular helix angle in a particular direction.
2. The cutting action is reciprocating, and cutting takes place only during one half of the stroke, so only about half the machine time is spent in metal removal.

Gear hobbing: Hobbing is the process of generating gear teeth by means of a rotating cutter called a "hob". A hob resembles a worm, with edges made parallel to its axis. Relief is provided behind each of the helically arranged cutting faces. Gear hobbing is a continuous cutting operation. The hob and the gear blank are connected by means of proper change gears. The ratio of the hob and blank speeds is such that during one revolution of the hob, the blank turns through as many teeth as there are starts (threads) as the hob.

To start cutting a thread, the hob is made to clean the blank. It is then moved inwards to obtain the required tooth depth. After the tooth depth is reached, the hob is fed in a direction parallel with the axis of rotation of the gear (axial hobbing). As the gear blank rotation, the teeth are generated and the feed of the hob across the face of the blank extends the teeth to the desired face width. One rotation of the blank completes the cutting unless the blank has a wide face.

Advantages of Gear Hobbing

1. The method produces accurate gears and is suitable for medium and large batch production.
2. The method is versatile and can generate spur, helical, worm and worm wheels.
3. The cutter is universal, because it can cut all gears of same module, irrespective of number of teeth on the gear.
4. Since gear hobbing is a continuous process, it is rapid, economical and highly productive.

Disadvantages

1. Gear hobbing cannot generate internal gears.
2. Enough space has to be there in component configuration for hob approach.

Applications

The gears produced by gear hobbing are used in automobiles, machine tools, various instruments, clocks and other equipments.

10. What conditions must be satisfied in order so that a pair of spur gears may have a constant velocity ratio?
SOLUTION: For constant velocity ratio the conjugate action for teeth profile must be satisfied. For conjugate action the pitch point must remain fixed. That is all the

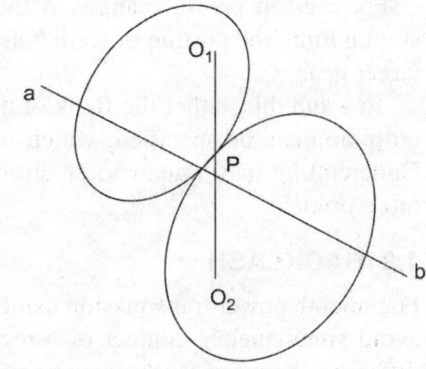

lines of action for every instantaneous point of contact must pass through the same point *P* as shown below. It is possible either by involute profile or centroidal profiles. In case of involute profiles all point of contacts occur on the same straight line *ab*. All normals to the tooth profiles at point of contact coincide with the line *ab*.

1.8 INTERFERENCE IN INVOLUTE GEARS

If addendum of gear (driven gear) interferes with the base circle of pinion or passes outside the path of contact and vice versa, this results in the removal of gear material from base circle (undercutting) and the phenomenon is called interference. Minimum number of teeth to avoid interference and undercutting are: 32, 17 and 14 respectively for 14.5° full depth, 20° full depth and 20° stub involute gears.

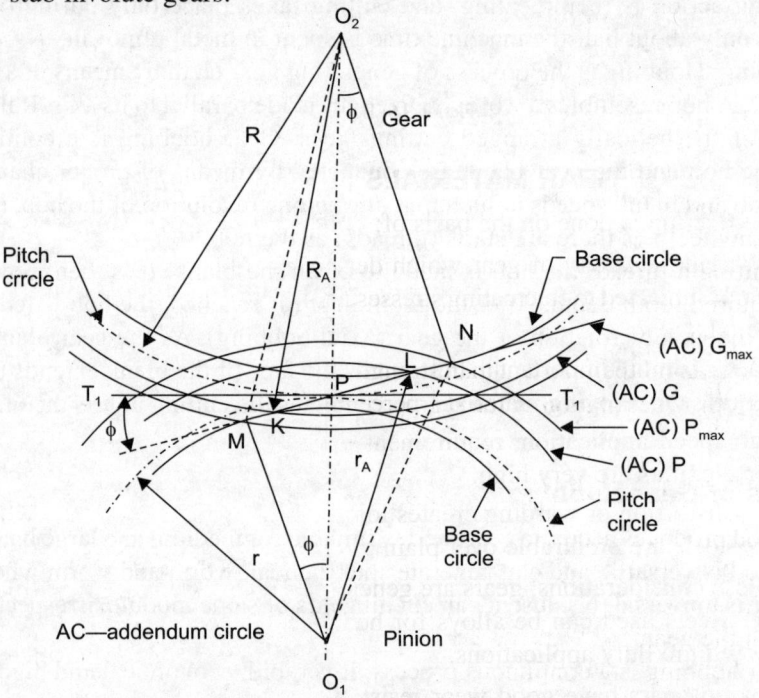

Fig. 1.15 Interference in involute gears

Interference can be avoided by—increasing the number of teeth on the pinion, increasing pressure angle or using long and short addendum gearing.

For a given pressure angle, if the number of teeth on the pinion are decreased below a certain limit, the portion of teeth below the base circle will interfere with the addendum of the larger gear.

To avoid this, either the flank of the pinion can be **undercut**, or minimum number of teeth on pinion can be specified, which mesh any gear without interference (as specified above). Undercutting may impair tooth strength significantly thereby opting for 2nd option may be more prudent.

1.9 BACKLASH

For smooth power transmission using gear drives, it is required to avoid jamming of teeth or to avoid simultaneous contact of a teeth at 2 or more points and is numerically equal to the difference between tooth space width along pitch circle and tooth space.

Backlash also compensates for some machinery errors and thermal expansion introduced by slight increase in center distance between shafts and slightly thin teeth on gears.

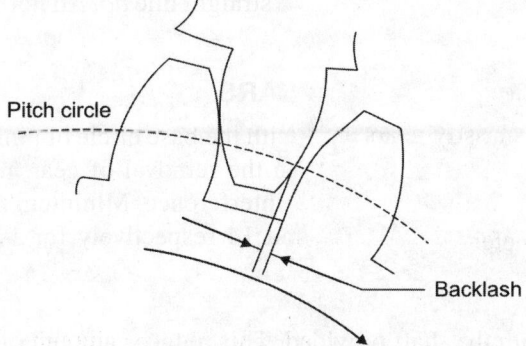

Fig. 1.16 Backlash

1.10 SELECTION OF GEAR MATERIALS

Gear material selection is done on the basis of:

a. Load carrying capacity of gear which depends on ultimate tensile strength and yield strength. If subjected to fluctuating stresses, endurance strength is the criteria for selection of material.

b. Wear rating is more important than strength rating in deciding the dimensions. It depends upon grain size, alloying materials, percentage of carbon and surface hardness.

c. For high speed application, requirement of low coefficient of friction is required, as sliding velocities are very high.

d. Thermal distortion or warping creates load concentration on any corner of gear tooth, so alloy steels are preferable over plain carbon steels.

Based on these considerations, gears are generally made up of cast iron, steels, bronze and phenolic resin. Steels used can be alloys for heavy duty applications and plain carbon like $50C_8$, etc. for medium duty applications.

- Large sized gears have good wear resistance, but low load carrying capacity or low strength.

- In case of non-metallic gears, pinion (driving gear) is usually made up of non-metals like moulded nylon, phenolics, etc. whereas gear (driven part) is made up of cast iron.

- Case hardened steel gears have good wear resistant surface together with ductile and shock absorbing core.

- Phenolic resins as gear material have low modulus of elasticity and work on marginal lubrication and can tolerate errors in tooth profile.

- Bronze is used in worm wheels due to low coefficient of friction and excellent formability. Bronze is used as gear material for following cases:
  o When long gear life is required.
  o When gears are likely to be affected by water and oil.
  o When low loads and less pitch line velocity (circumferential velocity of gear).

The only drawback being excessive cost while using bronze as gear material.

1.11 GEAR MANUFACTURING METHODS

These mainly include:

- Machining
- Casting
- Forming

Manufacturing methods classify gears as:

- Small gears
- Medium sized gears
- Gears with large diameter

1.11.1 Small Gears

Gear is made integral with the shaft provided. This reduces amount of machining and number of parts like keys (for securing gear to shaft). Also shaft rigidity and accuracy of contact are increased.

For $(d_d - d_s) < d_s/2$ (gear is considered as small)

where d_d = Diameter of dedendum

       d_s = Shaft diameter

then d $= mz$

           $d_a = mz + 2m = m(z + 2)$

           $d_d = mz - 2.5m = m(z - 2.5)$

where m = gear module

       z = Number of gear teeth

       d_a = Addendum diameter

       d_d = Dedendum diameter

In figure,

       b = Face width = Width of gear

       d_f = Dedendum circle diameter

       d' = Pitch circle diameter

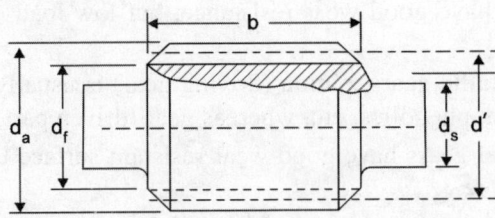

Fig. 1.17 Integral gear

1.11.2 Medium Sized Gears

Manufacturing of medium sized gears is done by:

- Machining from rolled steel bars, i.e. gear blanks obtained by turning on lathe. Usually followed for gear diameter less than 150 mm.
- Forging in open or closed dies. Usually followed for gear diameter lying in between the range 150–400 mm.
- Connecting hub, web and rim.

Forging has following advantages in the manufacture of medium sized gears:

a. Factor of utilization is equal to 1/3 in machining whereas it is 2/3 in forging.

b. Forged gear is of light weight construction which reduces inertial and centrifugal forces.

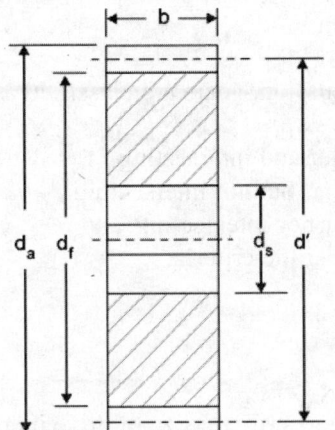

Fig. 1.18 (a) Machined gear

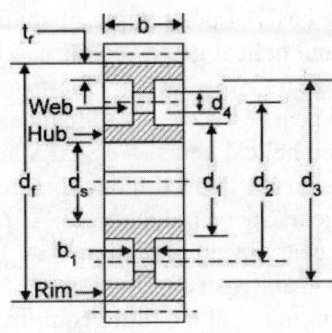

Fig. 1.18 (b) Forged gear

c. Fiber lines of forged gear suit the direction of external force whereas in case of machined gears fiber lines may be broken thereby making forged gears more strong.

1.11.3 Gears with Large Diameter

Large diameter gears are manufactured as:

- Solid cast gears
- Rimmed gears

1.11.3.1 Solid cast gears

For addendum diameter ≤ 900 mm, gears are manufactured as solid cast iron with one web and for addendum diameter $> 1,000$ mm, gear is manufactured as solid cast iron with 2 webs.

1.11.3.2 Rimmed gears

Gear rim is forged from alloy steel so as to provide high strength at economical cost.

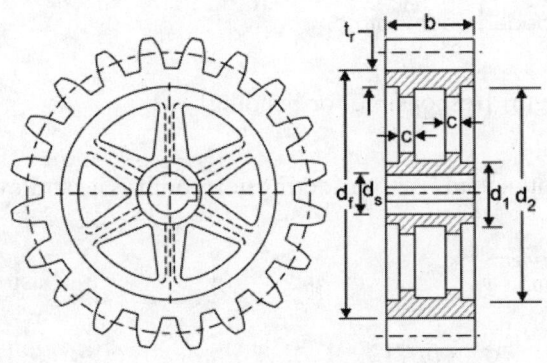

Fig. 1.19 Cast iron web type gear

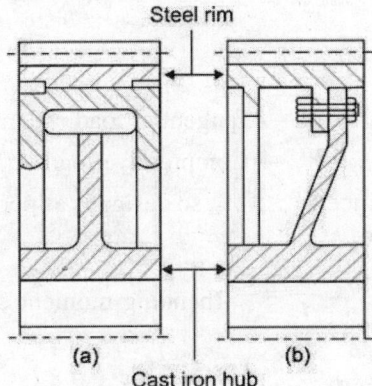

Fig. 1.20 Rimmed gear

1.12 DESIGN CONSIDERATIONS

Factors to be considered for effective gear design are:
a. Power to be transmitted using gear drive
b. Speed reduction required
c. Input speed and cost
d. General layout of shafts
   On the basis of positioning of shafts for power transmission, gears are classified as:
   Spur and helical gears – For parallel shafts
   Bevel gears   – Shafts axes are perpendicular and intersecting
   Worm gears   – Shafts axes are perpendicular, but not intersecting
   Crossed helical gears – Shafts neither perpendicular nor intersecting
   On the basis of speed reduction, gears are chosen as per the criteria:
   Spur gear: Speed reduction   – 6:1 to 10:1
   Bevel gear: Speed reduction   – 1:1 to 3:1
   Worm gears: Speed reduction   – 60:1 to 100:1
   (used in material handling equipment)
   - For increase in speed reduction requirement using gear drive, size of drive increases so as to maintain compact drive to increase speed 2 or 3 stages reduction is done.
   - For high speed power transmission, helical gears are usually preferred.

1.13 BEAM STRENGTH OF GEAR TOOTH

This is the maximum value of the tangential force that the tooth can transmit without bending failure.

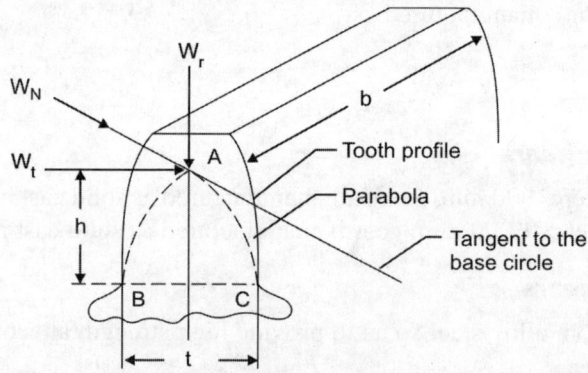

Fig. 1.21 Loaded gear tooth

here

$W_t \rightarrow$ Tangential load (beam strength) [responsible for bending]

$W_r \rightarrow$ Compressive load

Since $W_t >> W_r$, so design is as per tangential load W_t. Considering gear tooth as a cantilever beam:

$$\frac{M}{I} = \frac{\sigma}{y} \text{ (bending moment equation)}$$

$$\Rightarrow \quad \sigma = \frac{M \times y}{I} \Rightarrow M = \frac{\sigma \times I}{y} \Rightarrow W_t \times h = \frac{\sigma_b \times bt^3/12}{t/2} \qquad \text{(substituting } I \text{ and } y\text{)}$$

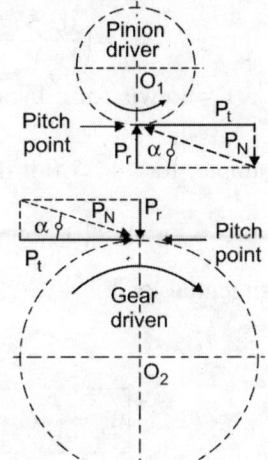

Fig. 1.22 Gear tooth force (gears in mesh)   **Fig. 1.23** Components of tooth force

b = Tooth face width

t = Tooth thickness

$I = bt^3/12$, $y = t/2$ and $\sigma = \sigma_b$ (from gear tooth geometry)

$\Rightarrow$ $W_t = \dfrac{\sigma_b bt^2}{6h} = \sigma_b bm\left(\dfrac{t_2}{6hm}\right)$ [multiplying and dividing by module 'm']

taking $\dfrac{t^2}{6hm} = Y$ (Lewis form factor)

$\Rightarrow$ $\boxed{W_t = \sigma_b\ b\ m\ Y}$ $[\sigma_b$ = Permissible bending stress $= \dfrac{1}{3}(S_{ut})$

S_{ut} = Ultimate tensile stress]

here W_t = Beam strength of gear tooth responsible for bending

1.14 DYNAMIC TOOTH LOAD

For involving dynamic effect of load:

$$\boxed{W_t = \sigma_b C_V b\ m\ Y}$$

C_V = Coefficient of velocity

$= \dfrac{3}{3+V}$(for $V < 8$ m/sec) $\left[V = \text{Pitch velocity} = \dfrac{\pi DN}{60}\right]$

$= \dfrac{4.5}{4.5+V}$(for $V < 13$ m/sec) N—rpm, D—diameter

$= \dfrac{6}{6+V}$(for $V < 20$ m/sec)

$= \dfrac{5.5}{5.5+\sqrt{V}}$(for $V > 20$ m/sec)

$$= \left(\frac{0.75}{1+V} + 0.25 \right) \text{(for non-metallic gears)}$$

Y = Lewis form factor (value can be seen from Design Data Handbook for various gear tooth systems.)

for example, for $14^1/_2°$ full depth involute system:

$$Y = \left(0.124 - \frac{0.684}{T} \right) \text{ [T = Number of teeth]}$$

also tangential load:

$$\boxed{W_t = \frac{P}{V} C_S}$$

$$\left[\begin{array}{l} P = \text{Power in watts} \\ V = \text{Pitch line velocity} \\ C_S = \text{Service factor} = \dfrac{\text{Maximum torque}}{\text{Rated torque}} \end{array} \right]$$

So on incorporating dynamic loads,

$$\boxed{W_d = C_S W_t + W_i}$$

$[W_d$ = Dynamic loads

W_i = Incremental load]

also

$$\boxed{W_d = \frac{C_S}{C_V} \times W_t} = P_{eff}$$

$$\boxed{W_d = \frac{P}{V} + \frac{21V(bC + W_t)}{21V + \sqrt{bC + W_t}}} = P_{eff}$$

here C = Dynamic deformation factor in N/mm^2

$$= \frac{K'}{\left(\dfrac{1}{E_p} + \dfrac{1}{E_g} \right)}$$

[K' = Constant depending upon tooth form

= 0.107 for 14 1/2° FD

0.111 for 20° FD

0.115 for 20° stub involute]

E_p and E_g being modulus of elasticity for pinion (driving gear) and gear (driven).

Surface endurance strength: $\boxed{W_S = \sigma_e bmY}$

here σ_e = Endurance strength

To avoid breakage of gear teeth:

$$\boxed{W_S > W_d}$$

1.15 TOOTH WEAR LOAD/WEAR STRENGTH OF GEAR TOOTH

$$W_w = d_p \, bQK$$

(W_w = Wear load, d_p = Pinion diameter, b = Face width, Q = Ratio factor, K = Load stress factor)

$$b = (9 \text{ to } 12) \text{ times module} = (9 - 12) \times m$$

$$Q = \frac{2T_g}{T_g + T_p} \text{ for external gears} \quad [T_g = \text{Number of teeth on driven gear,}$$

T_p = Number of teeth on pinion or driving gear]

$$= \frac{2T_g}{T_g - T_p} \text{ for internal gears.}$$

$$K = \frac{(\sigma_{es})^2 \sin\phi\cos\phi}{1\cdot4}\left[\frac{1}{E_p} + \frac{1}{E_g}\right] = 0.16\left(\frac{BHN}{100}\right)^2 \qquad [\text{for } 20° \text{ full depth involute}]$$

Wear strength is the maximum value of the tangential force that the tooth can transmit without pitting failure.

1.16 FAILURE OF GEAR TOOTH

Failure modes of gear tooth are:
• Static and dynamic loads
• Surface destruction

Static and dynamic loads can be avoided by adjusting module and face width such that the sum of static and dynamic loads is less than beam strength of gear tooth.

Primary causes of surface destruction are:

• Abrasive wear: Foreign particles in lubricant like dirt, rust, weld spatter, metallic debris, etc. can scratch the surface and is called abrasive wear.

  How to avoid: Provision of oil filters, increasing surface hardness and use of high viscosity oils allow fine particles to pass without scratching.

• Destructive pitting: This is a surface fatigue failure occurring due to load on tooth exceeding surface endurance strength or wear strength of materials and depends on Hertz contact stresses and number of stress cycles. Destructive pitting is subsequently characterized by pits which continue to grow resulting in complete destruction of tooth surface.

  How to avoid: Gear designing in such a way that wear strength of gear tooth is greater than the sum of static and dynamic loads and improving surface endurance strength by increasing surface hardness.

• Scoring: Excessive surface pressure, high surface speed and inadequate supply of lubricant result in excessive frictional heating of gear tooth. Scoring is a stick-slip phenomenon characterized by alternate welding and shearing rapidly at high speeds and rate of wear is faster.

  How to avoid: Bulk temperature of lubricant can be reduced by providing fins on outside surface of gear box and a fan for forced circulations of air over the fins.

• Corrosive wear: Due to presence of extreme pressure additives in lubricating oils and external contamination resulting in wear which is uniformly distributed over the entire surface.

  How to avoid: Providing complete enclosure for gears free from external contamination, change of lubricant periodically and selection of proper additives.

• Initial pitting: Presence of small pits at high spots caused by errors in tooth profile, surface irregularities and misalignment.

  How to avoid:
  o Proper machining of gears
  o Precise alignment
  o Reducing dynamic loads

1.17 DESIGN OF SPUR GEARS

1.17.1 Estimation of Module Based on Beam Strength

$$S_b = \sigma_b\, bmY \qquad\qquad (Y = \text{Lewis form factor from Design Data Handbook})$$

For proper design of spur gear drive:

$$S_b \geq P_{eff}$$

$$\Rightarrow \qquad S_b = P_{eff} \times f_s \qquad\qquad [f_s = \text{Factor of safety} = 1.5\text{--}2]$$

So

$$\boxed{S_b = \frac{C_S}{C_V} P_t \times f_s} \qquad\qquad \text{...(A)}$$

Now

$$P_t = 2\frac{M_t}{d} = \frac{2}{d} \times \left(\frac{60 \times 10^6\, \text{kW}}{2\pi N} \right)$$

(as tangential load acts at pitch circle radius) $\left[\text{also } P = \dfrac{2\pi NM}{60} \right]$

$$\Rightarrow \qquad M = \frac{60 \times 10^6\, (\text{kW})}{2\pi N}$$

So

$$\boxed{P_t = \frac{2}{mT} \times \left(\frac{60 \times 10^6\, \text{kW}}{2\pi N} \right)} \qquad \left[\text{as } m = \text{Module} = \frac{d}{T} = \frac{\text{Diameter}}{\text{No. of teeth}} \right]$$

So

$$\sigma_b\, bm\, Y = \frac{C_S}{C_V}(f_s) \times \frac{2}{mT} \times \left(\frac{60 \times 10^6\, \text{kW}}{2\pi N} \right)$$

$$\Rightarrow \qquad \left(\frac{S_{ut}}{3} \right)\left(\frac{b}{m} \right) m^2 Y = \frac{C_S}{C_V}\left(\frac{60 \times 10^6\, \text{kW}}{mT\pi N} \right) f_s$$

$$\Rightarrow \qquad m^3 = \left[\frac{(60 \times 10^6\, \text{kW})\, C_S f_s}{\left(\dfrac{S_{ut}}{3} \right)\left(\dfrac{b}{m} \right) C_V NTY} \right]$$

$$\Rightarrow \qquad m = \left[\left(\frac{60 \times 10^6}{\pi} \right)\left[\frac{(\text{kW})\, C_S f_s}{\left(\dfrac{S_{ut}}{3} \right)\left(\dfrac{b}{m} \right) C_V NTY} \right] \right]^{1/3} \qquad (\text{kW} = \text{Power in kilowatts})$$

1.17.2 Estimation of Module Based on Wear Strength

$$W_w \geq P_{eff} \qquad\qquad [W_w = \text{Tooth wear load}]$$

$$\Rightarrow \qquad d_p\, bQk = P_{eff} \times f_s$$

$$\Rightarrow \qquad d_p\, bQk = \frac{C_S}{C_V} P_t \times f_s$$

So

$$P_t = \frac{2M_t}{d} = \frac{2}{M_t} \times \left(\frac{60 \times 10^6\, (\text{kW})}{2\pi N} \right)$$

$$= \frac{60 \times 10^6\, (\text{kW})}{\pi N m T}$$

$$\Rightarrow \qquad d_p b Q k = \frac{C_S}{C_V} f_s \times \frac{60 \times 10^6 \, (\text{kW})}{\pi N m T}$$

$$\Rightarrow \qquad m T \times \left(\frac{b}{m}\right) \times m Q k = \frac{C_S}{C_V} f_s \left(\frac{60 \times 10^6 \, \text{kW}}{\pi N m T}\right)$$

$$\Rightarrow \qquad m^3 = \frac{C_S f_s \left(60 \times 10^6 \, (\text{kW})\right)}{C_V \left(\dfrac{b}{m}\right) Q k \times \pi N T^2}$$

$$\Rightarrow \qquad m = \left[\left(\frac{60 \times 10^6}{\pi}\right) \frac{C_S f_s \, (\text{kW})}{C_V \left(\dfrac{b}{m}\right) Q k N_p T_p^2}\right]^{1/3}$$

1.17.3 Design Criteria for Spur Gear

Calculate $\boxed{\sigma_b \times Y}$ for both pinion (driving gear) and driven gears. [$\sigma_b = \dfrac{1}{3} S_{ut}$ and

Y = Lewis form factor]

Whichever value is less, that gear will be weaker and so design will be as per the weaker component and module is calculated on the basis of beam strength or wear strength accordingly.

WORKED EXAMPLES

1. Design a spur gear and specify dimensions and suitable surface hardness for following set of gear parameters.

ϕ (pressure angle) = 20° full depth involute

P_p (power transmitted on pinion) = 10 kW

N_p = 1,440 rpm

C_S = Service factor = 1.5

Velocity ratio = V_r = 4:1

S_{ut} (ultimate tensile strength) = 600 N/mm^2, factor of safety = 1.5

SOLUTION: Let $b = 10 \, m$ (b is face width, m is module)

$$M_t = \frac{60 \times 10^6 \, P_{\text{kW}}}{2 \pi N} = \frac{60 \times 10^6 \times 10}{2 \pi \times 1440}$$
$$= 66{,}314.56 \text{ N-mm.}$$

also $\qquad V_r = \dfrac{V_p}{V_g} = \dfrac{T_g}{T_p} = \dfrac{4}{1}$

$\Rightarrow \qquad T_g = 4 T_p = 4 \times 18 = 72 \qquad$ (as minimum no. of teeth on pinion for 20° FD involute is 17, so taking $T_p = 18$)

Effective tooth load

$$P_{eff} = \frac{C_S}{C_V} \times P_t$$

assuming pitch line velocity (V) = 5 m/sec.

So $\qquad C_V = \dfrac{3}{3+V} = \dfrac{3}{3+5} = \dfrac{3}{8}$ *(from table $C_V = \dfrac{3}{3+V}$)

So $\qquad m = \dfrac{60 \times 10^6}{\pi} \left[\dfrac{P_{kW} \times C_S \times f_s}{T_p N_p C_V \left(\dfrac{b}{m}\right)\left(\dfrac{S_{ut}}{3}\right) Y} \right]^{1/3}$

* also from table for $T_p = 18$, $Y = 0.308$.

$\Rightarrow \qquad\qquad m = 4.16 \text{ mm} \approx 5 \text{ mm}$

Trial 1:

$\qquad m = 5 \text{ mm}$

$\qquad d_p = m\, T_p = 5 \times 18 = 90 \text{ mm}$

$\qquad d_g = m\, T_g = 5 \times 72 = 360 \text{ mm}$

$\qquad b = 10\, m = 10 \times 5 = 50 \text{ mm}$

Check for design:

$$P_{eff} = \dfrac{C_S}{C_V} P_t, \; P_t = \dfrac{2M_t}{d_p} \qquad\qquad \left[\text{as } M_t = P_t \times \dfrac{d_p}{2} \right]$$

$\Rightarrow \qquad\qquad P_t = \dfrac{2 \times 66{,}314.56}{90} = 1{,}473.66 \text{ N}$

So $\qquad\qquad P_{eff} = \left(\dfrac{1.5}{3/8}\right) \times 1{,}473.66 = 7{,}209.69 \text{ N}$

$\qquad\qquad S_b = \sigma_b\, bmY \quad \text{(beam strength)}$

$$= \left(\dfrac{S_{ut}}{3}\right) bmY = \left(\dfrac{600}{3}\right) \times 50 \times 5 \times 0.308$$

$$= 15{,}400 \text{ N}$$

also $\qquad\qquad S_b = P_{eff} \times (f_s) \Rightarrow 15{,}400 = 7{,}209.69 \times f_s$

$\Rightarrow \qquad\qquad f_s = \dfrac{15{,}400}{9{,}209.69} = 2.14$, so design is satisfactory.

Surface hardness:

$$S_w = d_p\, bQk = P_{eff} \times f_s$$

$$= 18 \times 50 \times \left(\dfrac{2T_g}{T_g + T_p}\right) \times 0.16 \times \left(\dfrac{BHN}{100}\right)^2 \qquad \text{[for 20° FD involute]}$$

$\Rightarrow \qquad\qquad S_w = 900 \times \left(\dfrac{2 \times 72}{72 + 18}\right) \times 0.16 \times \left(\dfrac{BHN}{100}\right)^2$

$$= 7{,}209.69 \times 15$$

$\qquad\qquad BHN = 306.4$

2. For given set of parameters, design a spur gear (determine the module).

$\qquad T_p$ = Number of teeth on pinion = 20

$\qquad T_g$ = Number of teeth on driver gear = 50

P_p = Power transmitted on pinion = 22.5 kW

N_p = 1450 rpm

$$C_S = \frac{150}{100} = 1.5 = 1.5 \left(\frac{\text{Operating torque}}{\text{Rated torque}} \right)$$

S_{ut_p} = Ultimate tensile strength (plain carbon steel)

= 420 N/mm^2 (as pinion material)

S_{ut_g} = Ultimate tensile strength (grey cast iron)

= 200 N/mm^2 (as driven gear material)

f_s = 1.5. (factor of safety)

SOLUTION: As per the design criteria of spur gears, value of ($\sigma_b \times Y$) is calculated for both pinion and gear. Whichever is less, design will be done as per the weaker component (pinion or driven gear)

For pinion: $\quad \sigma_b \times Y = \left(\dfrac{S_{ut_p}}{3} \right) \times Y = \left(\dfrac{410}{3} \right) \times 0.320 = 43.73$

For gear: $\quad \sigma_b \times Y = \left(\dfrac{S_{ut_g}}{3} \right) \times Y = \left(\dfrac{200}{3} \right) \times 0.408 = 27.20$

So gear is weaker and so design will be as per the weaker component.

So $\quad m = \left[\dfrac{60 \times 10^6}{\pi} \left(\dfrac{22.5 \times 1.5 \times 1.5}{N_g T_g C_V Y_g \left(\dfrac{b}{m} \right) \left(\dfrac{S_{ut}}{3} \right)} \right) \right]^{1/3}$

$\quad = \left[\dfrac{60 \times 10^6}{\pi} \left(\dfrac{22.5 \times 1.5 \times 1.5}{580 \times 50 \times \left(\dfrac{3}{8} \right) \times (0.408)(10) \times \left(\dfrac{200}{3} \right)} \right) \right]^{1/3} = 6.89 \approx 7\,\text{mm}$

as $\quad \dfrac{N_p}{N_g} = \dfrac{T_g}{T_p} \Rightarrow N_g = \dfrac{T_p}{T_g} \times N_p = \dfrac{20}{50} \times 1450$

So $\quad N_g = 580$

assuming $\quad V$ = Pitch line velocity = 5 m/sec.

So $\quad C_V = \dfrac{3}{3+V} = \dfrac{3}{3+5} = \dfrac{3}{8}\,\text{m/sec.}$

and $\quad b = 10m$

So $\quad b = 10 \times 7 = 70$ mm

$\Rightarrow \quad d_p = mT_p = 7 \times 20 = 140$ mm

$\quad d_g = mT_g = 7 \times 50 = 350$ mm

Test for design:

As beam strength for spur gear

$\quad S_b = P_{eff} \times$ Factor of safety

also $\quad S_b = \sigma_b\, bmY$

$\Rightarrow$ $\qquad S_b = \dfrac{S_{ut}}{3} \times b \times mY = \dfrac{200}{3} \times 70 \times 7 \times (0.408)$

$\Rightarrow$ $\qquad S_b = 13,328 \text{ N}$

also $\qquad P_{eff} = \dfrac{C_S}{C_V} \times P_t$

$\qquad V = \dfrac{\pi dN}{60} = \dfrac{\pi \times 350 \times 580}{60} = 10.63 \text{ m/sec.}$

So $\qquad C_V = \dfrac{6}{6+V} = 0.3$

$\Rightarrow$ $\qquad P_t = 2 \times \dfrac{M_t}{d}$ $\qquad$ [as $M_t = P_t \times d/2$]

So $\qquad P_t = \dfrac{2}{d_g} \times \dfrac{60 \times 10^6 (P_{kW})}{2\pi N_g}$

$\qquad\qquad = \dfrac{2}{350} \times \dfrac{60 \times 10^6 \times 22.5}{2\pi \times 580} = 2{,}116.84 \text{ N}$

So $\qquad P_{eff} = \dfrac{1.5}{0.3} \times 2{,}116.84 = 10{,}584.2$

So factor of safety $= \dfrac{S_b}{P_{eff}} = \dfrac{13{,}328}{105{,}842} = 1.26$

So design is safe and module = 7 mm

3. A pair of carefully cut gears with 20° involute teeth is to transmit 25 kW at 300 rpm of the gear at a speed reduction of 5:1. The pinion should not be smaller than 76 mm in pitch diameter and is made of forged C-30 steel with a hardness 250 BHN and drives cast-iron gear. Select a suitable spur gear and determine module, number of teeth and face width of the gears.

SOLUTION: Given: Pressure angle (ϕ) = 20°

$\qquad\qquad\qquad$ Power transmitted (P) = 25 × 10³ W

$\qquad\qquad\qquad\qquad$ Speed (N) = 300 rpm (gear speed)

$\qquad\qquad\qquad$ Reduction ratio = 5:1

$\qquad\qquad$ Pitch diameter of pinion = 76 mm

Pinion material is C-30 with hardness 250 BHN

$\qquad\qquad\qquad\qquad$ Gear material = Cast iron

Problem may be solved by first assuming minimum number of teeth necessary to avoid interference and then checking it at the end from the module obtained.

Let $T_p = 18$, so corresponding teeth on gear

$\qquad T_g = 90$

Applying Lewis equation for pinion and gear (for 20° full depth)

$\qquad y_p = 0.154 - \dfrac{0.912}{T_p} = 0.154 - \dfrac{0.912}{18} = 0.1033$

$\qquad y_g = 0.154 - \dfrac{0.912}{90} = 0.143867$

Material for pinion is C-30 steel (hardness 250 BHN).

From data table allowable static stress (σ_0) is MPa or N/mm^2.

For pinion it is 110 and for gear CI (cast iron driven gear) it is 55 MPa or N/mm^2.

$$\therefore \qquad \sigma_{op} \times y_p = 110 \times 0.1033 = 11.363 \qquad \qquad ...(1)$$

$$\sigma_{og} \times y_g = 55 \times 0.143867 = 7.9126 \qquad \qquad ...(2)$$

Comparing (1) and (2), we known that the value of $\sigma_{og} y_g$ is less than $\sigma_{op} y_p$. Therefore, driven gear is weaker. So it has to be designed.

Linear velocity $(V) = \dfrac{\pi d N}{60}$

Diameter of pinion $= 76$ mm

We know that module $(m) = \dfrac{d_g}{T_g} = \dfrac{d_p}{T_p}$

$$\therefore \qquad \dfrac{d_g}{90} = \dfrac{76}{18} \text{ or } d_g = 76 \times 5 = 380 \text{ mm}$$

So $\qquad V = \dfrac{\pi \times 380 \times 300}{60} = 5.96 \approx 6 \text{ m/sec.}$

For carefully cut gears operating at velocity up to 12.5 m/sec

$$K_V = \dfrac{4.5}{4.5 + V} \text{ or } K_V = \dfrac{4.5}{4.5 + 6}$$

$$K_V = 0.4286$$

Designs tangential tooth load

$$F_t = \dfrac{P}{V} \times C_S$$

For steady load condition and 8–10 hours of service per day, service factor (C_S) taken from Data book is 1.

$$\therefore \qquad F_t = \dfrac{25 \times 10^3}{6} = 4,166.7 \text{ N}$$

Let face width (b) = $14 m$ (Taking in middle range from Data book)

Lewis equation for the gear may be written as

$$F_t = K_v\, (\sigma_{og} y_g) b . P_c, \text{ (where } P_c = \text{circular pitch)}$$

$$\therefore \qquad 4,166.7 = 0.4286 \times 7.9126 \times 14m \times \pi m$$

Solving this: or $m^2 = 32.5 \Rightarrow m = 5.28$ mm say 6 mm

Selecting module (m) = 6 mm

Face width (b) = $14m = 14 \times 6 = 84$ mm

We know that module

$$m = d/T$$

Because pitch circle diameter of pinion is greater than 76 mm.

$\therefore \qquad$ For pinion $6 \geq 76/T_p$

$$T_p \geq 13$$

$$T_p = 14 \text{ (say)}$$

Corresponding to this combination, corresponding no. of tooth in gear

$$T_g = 14 \times 5 = 70$$

To avoid interference combination of $T_p = 18$ and $T_g = 90$ may be selected and in any condition pitch circle diameter of pinion will not be less than 76 mm.

PREVIOUS YEAR UNIVERSITY QUESTIONS

1. A pair of gears is to be designed to transmit 4 kW at 600 rpm of the pinion to a gear rotating at 150 rpm 20° FD involute teeth is to be used. The center distance should be as small as possible. Only from strength considerations (Lewis equations), determine the module, face width and number of teeth on gears assuming spur gear drive.

SOLUTION: Given:

Power transmitted $(P) = 4$ kW

Speed of pinion $(N_p) = 600$ rpm

Speed of gear $(N_g) = 150$ rpm

Pressure angle $(\phi) = 20°$ FD

$$\text{Gear ratio} = G = \frac{N_p}{N_g} = \frac{600}{150} = 4.$$

Nothing is mentioned about diameters, center distance, materials and pitch line velocity.

Since speed of pinion and gear is not very high, a moderate value of pitch line may be selected. As the pitch line velocity for medium velocity drives ranges from 3 to 15 m/sec, let us choose a velocity 10 m/sec. Suitable materials for pinion and gear may be taken as forged C-40 steel with ultimate strength 580 MPa and cast iron grade 35, heat treated with ultimate tensile strength of 350 MPa respectively.

Approximate pitch diameter of pinion

$$d_p = \frac{60 \times V}{\pi N_p} = \frac{60 \times 10}{\pi \times 600} \Rightarrow d_P = 0.318 \text{ m}$$

Design load:

Now useful transmitted load

$$F_t = \frac{4 \times 10^3}{10} = \frac{\rho}{V}$$

or $$F_t = 0.4 \text{ kN}$$

Selecting carefully cut gears value of velocity factor (speed 10 m/sec) $C_V = \dfrac{4.5}{4.5 + V}$

or $$C_V = \frac{4.5}{4.5 + 10} = 0.31$$

Considering power as electronic motor, considering 24 hours/day operation and for medium shock.

$$C_S = 1.50 \times 1.25 = 1.875$$

Design tangential load $(F_t) = 1.875 \times 0.4$

$$F_t = 0.75 \text{ kN}$$

Gear strength calculation (Lewis equation)

$$F_t = \frac{F_b \times F \times m \times yC_V}{K_f}$$

where m = Module, K_f = Stress concentration factor
F = Face width of tooth, Y = Form factor
Let's take $Y = 0.29$ and $F = 12m$
$K_f = 1.5$ (from Data Handbook)
considering factor of safety $f_s = 3$

Design stress $\quad F_b = \dfrac{580}{3} = 193.3 \text{ MPG}$

$\therefore \qquad 0.75 \times 10^3 = \dfrac{193.3 \times 10^6 \times 12m \times m \times 0.29 \times 0.31}{1.5}$

or $\qquad m^2 = 5.39 \times 10^{-6}$

Let us choose a standard value of 'm' as 6 mm.

Now number of teeth on pinion $= \dfrac{d_p}{m} = \dfrac{0.318}{6} \times 10^3$

or $\quad N_p = 53 > 17$ (minimum required)
Let $N_p = 54$, exact diameter (d_p) $= 54 \times m$
$= 54 \times 6 = 324$ mm

So $d_p = 0.324\,m$, $V = \pi dN = \pi \times 0.324 \times \dfrac{600}{60}$
$= 10.178$ m/sec

and $\quad C_V = \dfrac{4.5}{4.5 + 10.178} = 0.306$, $F_t = \dfrac{4 \times 10^3}{10.178} = 393 \text{ N}$

$F_t = 393 \times 10^{-3}$ kN, exact value of y for pinion $= \pi \left(0.154 - \dfrac{0.912}{N_p} \right)$

$$= \pi \left(0.154 - \dfrac{0.912}{54} \right)$$

$$y = 0.43$$

Putting this value in Lewis equation

$$F_t = \frac{F_b \times m \times F \times y \times C_V}{K_f}$$

$\therefore \qquad 393 = \dfrac{193 \times 10^6 \times 12\,m \times m \times 0.43 \times 10.178}{1.5}$

or $m = 2.4$ mm

Module $(m) = 3$ can be selected.

As modified, but 6 can be also selected.

Face width = $12\,m = 12 \times 6 = 72$ mm.

∴ Number of teeth on gear = $G \times N_p = 4 \times 54 = 216$

2. Design a spur gear drive required to transmit 45 kW at a pinion speed of 800 rpm. The velocity ratio is 3.5:1. The teeth are 20° full depth involute with 18 teeth on the pinion. Both the pinion and gear are made of steel.

SOLUTION: Given:

Power transmitted = 45 kW = 45×10^3 W

Pinion speed (N_p) = 800 rpm

Velocity ratio (V_r) = 3.5/1

So wheel speed $= \dfrac{800}{3.5} = 228.56$ rpm

Pressure angle (ϕ) = 20° full depth involute tooth

Number of teeth on pinion (T_p) = 18

Pinion and gear are made of steel.

Let m = Module in mm

d_p = Pitch circle diameter of pinion in mm.

We know that the pitch line velocity

$$V = \frac{\pi d_p N_p}{60} = \frac{\pi m T_p N_p}{60} = \frac{\pi \times m \times 18 \times 800}{60} = (0.7536\,m)\,\text{m/sec.}$$

Assuming steady load conditions and 8–10 hours of service per day, the service factor (C_S) from databook is taken as $C_S = 1$

We know that design tangential tooth load

$$W_t = \frac{P}{V} C_S = \frac{45 \times 10^3}{0.7536m} \times 1$$

$$W_t = \frac{59,713.4 \text{ N}}{m}$$

From Design Book, selecting relation for velocity factor (C_V) for ordinary cut gears operating at velocity up to 12.5 m/sec is

$$C_V = \frac{3}{3+V} = \left(\frac{3}{3+0.7536m}\right)$$

For 20° full depth involute teeth, the tooth form factor for the pinion is

$$y_p = 0.154 - \frac{0.912}{T_p} \Rightarrow y_p = 0.154 - \frac{0.912}{18} = 0.1033$$

Tooth form factor for the gear

$$y_g = 0.154 - \frac{0.912}{T_g}$$

$\Rightarrow \qquad T_g = T_p \times V_r = 18 \times 3.5 = 63$

Taking even no. of teeth $\Rightarrow T_g = 64$

∴ $$y_g = 0.154 - \frac{0.912}{64} = 0.13975$$

Material for both wheel and pinion is steel; selecting cast-steel untreated for which allowable static stress $(\sigma_0) = 140$ MPa

So $\qquad \sigma_{op} \times y_p = 140 \times 0.1033$

and $\qquad \sigma_{og} \times y_g = 140 \times 0.1337$

Since $(\sigma_{op} \times y_p)$ is less than $(\sigma_{og} \times y_g)$, therefore, the pinion is weaker using the Lewis equation to the pinion, we have

$$W_t = \sigma_{WP} \, b\pi m y_p = (\sigma_{op} \times C_V)b\pi m y_p$$

Taking width of teeth $(b) = 14m$

$$\Rightarrow \qquad \frac{59,713.4}{m} = 140 \times \left(\frac{3}{3+0.7536m}\right) \times 14m \times \pi m \times 0.1033$$

$$\Rightarrow \qquad \frac{59,713.4}{m} = \left(\frac{1907.25}{3+0.7536\,m}\right) \times m^2$$

$$31.308 \,(3 + 0.7536m) = m^3$$

or $\qquad 93.924 + 23.59m = m^3$

Solving this equation by hit and trial method, value of m $(6 < m < 7)$ is about 6.22 mm. The standard module is 8 mm, therefore, let us take $m = 8$ mm

Face width (b) = $14m$

So $b = 14 \times 8$ = 112 mm

Pitch circle diameter of the pinion and gear pitch circle diameter of pinion.

$$d_p = mT_p$$

or $\qquad d_p = 8 \times 18 = 144$ mm

or pitch circle diameter

$$D_g = mT_g = 8 \times 64 = 512 \text{ mm}$$

3. The pinion in a spur gear set having a speed ratio of 3 rotates at 900 rpm and has 20 full depth teeth of pressure angle 20°. The face width is 30 mm and the module is 3 mm. What is the power rating of the gear set based only on tooth bending strength, if both the gear and pinion are made of the same material having an ultimate strength of 400 MPa and an yield strength of 210 MPa? Assume a factor of safety of 2.

SOLUTION: Given: Pinion speed (N_p) = 900 rpm

$$\text{Velocity ratio} \,(V_r) = \frac{T_g}{T_p} = 3$$

For both (gear and pinion), yield strength = 210 MPa

Factor of safety = 2

$$\therefore \text{ Allowable stress} = \frac{210}{2} = 105 \,\text{MPa}$$

Pressure angle $(\phi) = 20°$ (full depth tooth)

So $\sigma_{og} = \sigma_{op} = 105$ MPa = 105 N/mm^2

Teeth on pinion $(T_p) = 20$.

Face width (b) = 30 mm, module (m) = 3 mm.

Pitch circle diameter of the pinion

$$d_p = mT_p = 3 \times 20 = 60 \text{ mm} = 0.06 \text{ m.}$$

Pitch line velocity $(V) = \dfrac{\pi d_p N_p}{60} = \dfrac{\pi \times 0.06 \times 900}{60}$

$$= 2.83 \text{ m/sec.}$$

Since pitch line velocity is less than 12.5 m/sec, therefore, velocity factor

$$C_V = \frac{3}{3+V} = \left(\frac{3}{3+2.83}\right) = 0.514$$

We know that for 20° full depth involute teeth, tooth form factor for the pinion

$$y_p = \left(0.154 - \frac{0.912}{T_p}\right)$$

$$= \left(0.154 - \frac{0.912}{20}\right)$$

or $\qquad y_p = 0.1084$

Tooth form factor for gear

$$y_g = \left(0.154 - \frac{0.912}{T_g}\right) = \left(0.154 - \frac{0.912}{3 \times 20}\right) = 0.139$$

$\sigma_{op} \times y_p = 105 \times 0.1084 = 11.382$

Since $(\sigma_{op} \times y_p)$ is less than $(\sigma_{og} \times y_g)$, therefore, pinion is weaker. Now using lower equation for the pinion, we have tangential load on the tooth (or beam strength of tooth).

$$W_t = \sigma_{op} \times b \times \pi \times m \times y_p$$

$$= (\sigma_{op} \times C_V)\, b\pi\, my_p$$

$$= (105 \times 0.514) \times 30 \times 3.14 \times 3 \times 0.1084$$

or $\qquad W_t = 1653.3 \text{ N.}$

Power that can be transmitted

$$P = W_t \times V$$

$$P = 1{,}653.3 \times 2.83$$

$$P = 4{,}679 \text{ W}$$

or $\qquad P = 4.68 \text{ kW}$ **Ans.**

4. A pair of 20° full depth straight teeth spur gears is to transmit 25 kW. The pinion rotates at 400 rpm and the velocity ratio is 1:4. The allowable static stresses of gear and pinion materials are 100 MPa and 120 MPa respectively. The pinion has 16 teeth and the face width is 12 times the module. Design gear for static strength.

SOLUTION: Pressure angle $(\phi) = 20°$ FD

Power transmit $(P) = 25$ kW

Pinion speed $(N_p) = 400$ rpm

Velocity ratio $(V_r) = 1{:}4$

Static allowable stresses

Gear: $(T_b)_g = 100$ MPa, pinion: $(T_b)_p = 120$ MPa

Number of teeth on the pinion $(T_p) = 16$

Face width $= 12 \times$ module $= 12m$

Design the gear for static strength

Number of teeth on the gear $(T_g) = V_r \times T_p$

$T_g = 4 \times 16 = 48$

Lewis form factor:

$$y = 0.154 - \frac{0.912}{T}$$

For pinion, $y_p = 0.154 - \frac{0.912}{16} = 0.097$

For gear, $\quad y_g = 0.154 - \frac{0.912}{48} = 0.135$

∴ $\quad (\sigma_b)_g y_g = 100 \times 0.135 = 13.5$

$(\sigma_b)_p y_p = 120 \times 0.097 = 11.64$

Since $\quad (\sigma_b)_g y_g > (\sigma_b)_p y_p$, hence pinion is weaker.

So the design will be based on pinion.

Let m = Module in mm.

Pitch circle diameter of pinion $(d_p) = mT_p$

$d_p = 16m$ (mm)

Peripheral velocity of pinion $(V) = \dfrac{\pi d_p N_p}{60}$

$$= \frac{\pi \times 16 \times m \times 400}{60} = 335.1032m \, (\text{mm/s})$$

$$V = 0.335m \, (\text{m/s})$$

Velocity factor assumed carefully at gears

$$C_V = \frac{4.58}{4.58 + V} = \frac{4.58}{4.58 + 0.335m}$$

Design tangential load

$$F_t = \frac{P \times C_S \times 1,000}{V}$$

$C_S = 1.80$ (assuming medium shock condition)

$$F_t = \frac{25 \times 1.80 \times 1,000}{0.335m} \, \text{N}$$

$$= \frac{134328.3582}{m} \, \text{N}$$

Face width $(b) = 12m$ (given)

Applying Lewis beam equation

$$F_t = \pi(\sigma_b)_p \, bym \, C_V$$

$$\frac{134,328.3582}{m} = \pi \times 120 \times 12m \times 0.097 \times m \times \frac{4.58}{4.58 + 0.335m}$$

$\Rightarrow 615,223.881 + 45,000m = 2,009.785m^3$

$\Rightarrow m^3 - 22.391 - 306.1143 = 0$

Solving by hit and trial method

$$m = 7.835$$

Standardize the module $(m) = 8$ mm.

∴ Pitch circle diameter $(d_p) = 16m = 16 \times 8 = 128$ mm.

Pitch line velocity

$$V = 0.335 \times m = 0.335 \times 8 = 2.68 \text{ m/sec}$$

$$F_t = \frac{134,328.3582}{8} \text{ N} = 16,791.045 \text{ N}$$

Face width $(b) = 12 \times m = 12 \times 8 = 96$ mm.

Beam strength of designed gear teeth

$$S_b = \pi(\sigma_b) \times b \times my$$

$$= \pi \times 120 \times 96 \times 8 \times 0.097 = 28,084.3304 \text{ N}$$

Since $S_b > F_t$. Hence design is successful.

Designed parameter module

$$m = 8 \text{ mm}$$

Gear pitch circle diameter $= 8 \times 48 = 384$ mm.

Pinion pitch circle diameter $= 8 \times 16 = 128$ mm.

Face width $(b) = 12 \times 8 = 96$ mm.

2

Helical Gears

2.1 INTRODUCTION

Helical gears are used for power transmission between parallel shafts and provide a smooth drive with a high efficiency of transmission. Helical gears involve gradual engagement and continuous contact of engaging teeth.

These involve transmission of radial loads and thrust loads among the engaging shafts and find application in:
- Rolling mills
- Steam turbine reducing gears
- Automobile timing gears, etc.

Gear teeth in helical gears are included at an angle called helix angle (in contrast to spur gears, having parallel to axis teeth), and are also used for power transmission between parallel shafts (like spur gears).

2.2 TERMINOLOGY

Helical gears consist of teeth cut on a cylinder, but at an angle to the axis of rotation. The helix formed by the teeth may be right-handed on one gear and left-handed on the other. Teeth of helical gears have line contact.

Helical gears need better lubrication and accurate machining and increased sliding action for quiet and smooth operation, whereas the presence of axial thrust requires provision of thrust bearing resulting in costly type of construction. Some additional terms (or advancement over spur gear) are:

2.2.1 Helix Angle and Face Width

For smooth and continuous operation, circular pitch (or linear distance between gear teeth) is given as:

$$P_c = b \tan \alpha$$

where P_c = Circular pitch, b = Face width and α = Helix angle (angle between tooth inclination (center line of gear tooth) and gear axis)

also $\qquad W_a = W_t \tan \alpha$

where $\qquad W_a$ = Axial thrust and W_t = Tangential load

So with increase in 'α', axial thrust increases which is undesirable.

This puts a limiting value over helix angle and this ranges from 20–35°.

Also, with decrease in face width, helix angle needs to be increased to achieve desired helical tooth action.

Face width (for helical gears) is given by:

$$b = \frac{1.15 \pi m}{\tan \alpha}, \text{where '}m\text{' is the module.}$$

For helical gears, the face width may vary between 12.5 m to 20 m and in terms of diameter of pinion (driving gears) 'd_p', varies between $1.5 d_p$ to $2 d_p$.

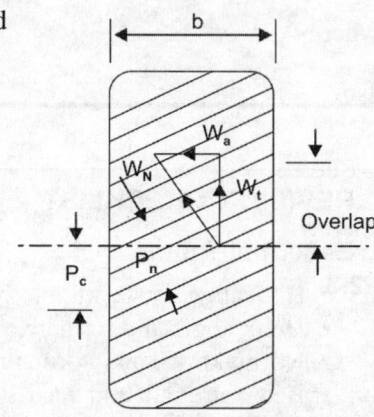

Fig. 2.1 Helical gear nomenclature

2.2.2 Formative Number of Teeth

Equivalent/formative number of teeth in helical gears, is the number of teeth which can be generated as a cylinder surface, with a radius equal to radius of curvature at tip of minor axis of an ellipse, obtained by analysis on a gear section in normal plane.

Numerically, T_E (formative no. of teeth) = $T/\cos^3 \alpha$

where 'T' is actual no. of teeth on helical gear.

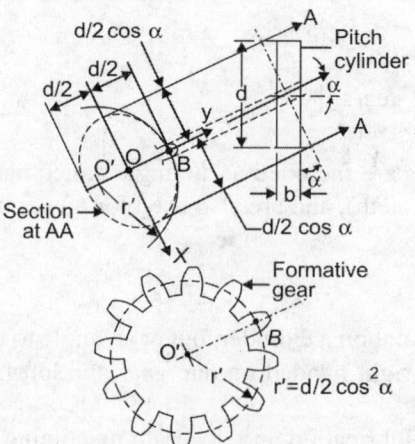

Fig. 2.2 Formative gear

To achieve this, an imaginary spur gear is considered in plane *A–A* with centre at O′ having a pitch circle radius of *r′* and normal module (m_n) (see next term). This is called formative gear or virtual spur gear.

The pitch circle diameter of the formative gear is given as: $d' = 2r' = \dfrac{d}{2\cos^2 \alpha}$, the number of teeth *T′*. On this virtual or formative gear is called virtual or formative number of teeth given by:

$$T' = \frac{2\pi r'}{P_n} = \frac{2\pi\left(d/2\cos^2 \alpha\right)}{\pi m_n} = \frac{d}{m_n \cos^2 \alpha} \qquad \text{...(1)}$$

(where P_n is the normal pitch and m_n is normal module (see next term))

also $\qquad\qquad d = \dfrac{Tm_n}{\cos \alpha}$ (see next term) $\qquad\qquad$ ...(2)

Substituting (2) in (1) $\Rightarrow T' = \dfrac{T}{\cos^3 \alpha}$

where *T* is the actual number of teeth.

2.2.3 Normal Pitch and Normal Module

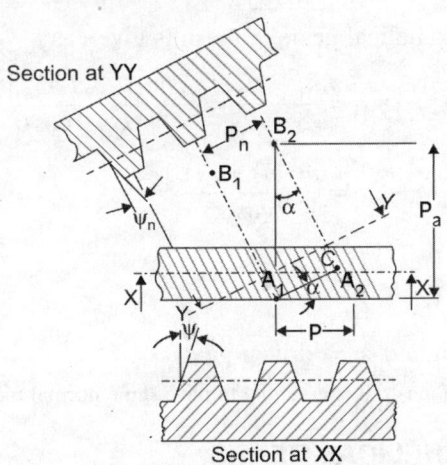

Fig. 2.3 Normal pitch and module

As shown in Fig. 2.3, lines A_1B_1 and A_2B_2 are centre lines of the adjacent teeth along the pitch plane. Angle '$A_1B_2A_2$' is the helix angle α. '*XX*' being rotation plane, while '*YY*' being plane perpendicular to the tooth element.

Distance 'A_1A_2' is called transverse circular pitch (*P*), measured in the plane of rotation.

Here distance 'A_1C' is called **normal circular pitch** (P_n) which is measured in a plane perpendicular to the tooth elements.

So from the figure: In $\Delta\, A_1A_2C$

$$\frac{P_n}{P} = \frac{A_1C}{A_1A_2} = \cos \alpha$$

or $\qquad\qquad \boxed{P_n = P \cos \alpha}$

also since $\qquad$ $P = \pi m \rightarrow$ (transverse module)

$\qquad\qquad\qquad P_n = \pi m_n \rightarrow$ (normal module)

$\Rightarrow$ $\qquad\qquad\qquad\qquad \boxed{m_n = m \cos \alpha}$

Distance $A_1 B_2$ is called the axial pitch (P_a).

From figure, in $\Delta A_1 A_2 B_2$, $\boxed{P_a = \dfrac{P}{\tan \alpha}}$

Similarly, for helical gears there are two pressure angles $\rightarrow$ transverse (ψ) and normal (ψ_n) pressure angles, which are related by:

$$\boxed{\cos \alpha = \frac{\tan \psi_n}{\tan \psi}}$$

Normal pressure angle ψ_n is usually 20°.

Pitch circle diameter (PCD) $\rightarrow$ (d) for helical gear is given as

$$d = \frac{TP}{\pi} = Tm = \frac{Tm_n}{\cos \alpha}$$

So $\qquad\qquad\qquad\qquad \boxed{d = \dfrac{Tm_n}{\cos \alpha}}$

Center distance between two helical gears in mesh is given as

$$a = \frac{d_1}{2} + \frac{d_2}{2} = \frac{T_1 m_n}{2 \cos \alpha} + \frac{T_2 m_n}{2 \cos \alpha}$$

or $\qquad\qquad\qquad\qquad a = \dfrac{m_n (T_1 + T_2)}{2 \cos \alpha}$

and speed ratio $= i = \dfrac{N_p}{N_g} = \dfrac{W_p}{W_g} = \dfrac{T_g}{T_p}$

where $\qquad\qquad p \Rightarrow$ pinion and $g \Rightarrow$ driven gear

*(**Note:** For two helical gears in mesh, it is essential to have same normal module 'm_n'.)

2.3 PROPERTIES FOR HELICAL GEARS

Helical gears are usually made as stub teeth. Pinion and gear are not interchangeable. Some standard helical gear tooth proportions are:

$\qquad$ Pressure angle $(\phi) = 15°$ to $25°$

$\qquad$ h_a = Maximum addendum = **$0.8 m_n$ or $1 m_n$**

$\qquad$ h_d = Minimum dedendum = **$1.0 m_n$ or $1.25 m_n$**

$\qquad$ Minimum depth = $1.8 m_n$

$\qquad$ Minimum clearance = **$0.2 m_n$ or $0.25 m_n$**

$\qquad$ Tooth thickness = $1.571 m_n$

$\qquad$ Addendum circle diameter $(d_a) = d + 2 h_a$

$$\boxed{d_a = \frac{Tm_n}{\cos \alpha} + 2 m_n = m_n \left(\frac{T}{\cos \alpha} + 2 \right)}$$

Dedendum circle diameter $(d_d) = d - 2h_d = \dfrac{Tm_n}{\cos\alpha} - 2.5m_n$

$$d_d = m_n\left(\dfrac{T}{\cos\alpha} - 2.5\right)$$

Normal pressure angle ψ_n is always $20°$
and helix angle (α) varies from $15°$ to $25°$.

Face width $\boxed{b \ge \dfrac{\pi m_n}{\sin\alpha}}$

2.3.1 Beam Strength of Helical Gears

Beam strength of helical gear is given by:

$\boxed{S_b = m_n b\sigma_b Y'}$ $\qquad$ S_b = Beam strength

m_n = Normal module

y' = Lewis form factor based on virtual or formative no. of teeth T'.

Beam strength (S_b) indicates maximum tangential force transmitted by gear tooth without bending failure.

*For proper design, it is essential that:
Beam strength > Efective force (P_{eff}) between meshing teeth.

or $\qquad$ Beam stress $= P_{eff} \times f_s$ (f_s = Factor of safety)

2.3.2 Wear Strength of Helical Gears

Wear strength of helical gears is given by:

$\boxed{S_w = \dfrac{bQd_pK}{\cos^2\alpha}} \rightarrow$ (Buckingham's equation of wear strength)

α = Helix angle, Q = Ratio factor $= \dfrac{2T_g}{T_g + T_p}$ (for external gears)

$$= \dfrac{2T_g}{T_g - T_p} \text{ (for internal gears)}$$

$$S_b = m_n b\ \sigma_b\ Y$$

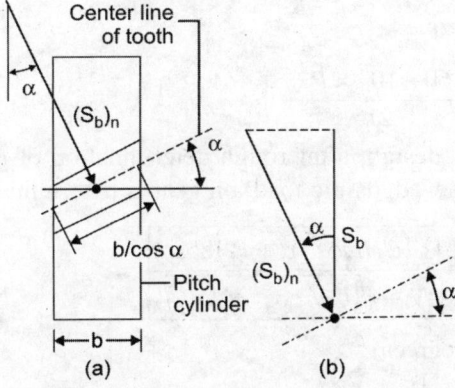

Fig. 2.4 Beam strength for helical gears

$$K = \frac{\sigma_c^2 \sin \psi_n \cos \psi_n \left(\dfrac{1}{E_p} + \dfrac{1}{E_g} \right)}{1.4}$$

σ_c = Surface endurance strength (N/mm^2)

E_p and E_g being modulus of elasticity for pinion and gear respectively.

ψ_n = Normal pressure angle

For steel gears with 20° normal pressure angle (ψ_n)

'K' modifies to $K = 0.16 \left(\dfrac{BHN}{100} \right)^2$

Wear strength (S_w) indicates the maximum tangential force that the tooth can transmit without pitting failure.

For proper design, wear strength (S_w) should be more than the effective force between the meshing teeth.

2.4 DESIGN OF HELICAL GEAR

For design of helical gears, effective load on gear tooth needs to be determined, both for beam strength as well as wear strength as design criteria.

• In the preliminary stages of design, effective tooth load 'P_{eff}' between 2 teeth in mesh is given by:

$$P_{eff} = \frac{C_S P_t}{C_V}$$

where $\qquad C_S$ = Service factor = Operating torque

and $\qquad C_V$ = Velocity factor $= \dfrac{5.6}{5.6 + \sqrt{V}}$ (for helical gears)

where V = Peripheral velocity of gear $= \left(\dfrac{\pi d N}{60} \right)$

(d = Pitch circle diameter, N = rpm)

and P_t = Tangential component of the resultant force between two meshing teeth of helical gears.

and is obtained as $\rightarrow P_t = \dfrac{2M_t}{d}$ (d = Pitch circle diameter)

where M_t = Transmitted torque

$$= \left(\frac{60 \times 10^6 \times P_{kW}}{2\pi N} \right) \qquad \text{[}P_{kW} \text{ = Power in kilowatt]}$$

• In final stages of gear design, after rough determination of gear dimensions, quality of gears and error specification, 'dynamic load' on gear is determined by following equation:

$$\boxed{P_d = \frac{21V \left(Ceb\cos^2 \alpha + P_t \right) \cos \alpha}{21V + \sqrt{\left(Ceb\cos^2 \alpha + P_t \right)}}}$$

α = Helix angle (degrees)

P_d = Dynamic load (N)

V = Pitch line velocity (m/s)

C = Deformation factor (N/mm²)

[depends upon modulus of elasticity and pressure angle]

b = Face width of tooth (mm)

P_t = Tangential force due to rated torque

e = Sum of errors between meshing teeth = $e_p + e_g$

e_p and e_g are obtained as per gear grade table,

for example, for grade 4 gear, $e = 3.20 + 0.25\,\phi$ and $\phi = m_n + 0.25\sqrt{d}$

Effective load is then given by (d is the pitch circle diameter for pinion and gear respectively)

$$\boxed{P_{eff} = C_S P_t + P_d}$$

• Now, to avoid failure of gear tooth due to bending: Beam strength: $S_b > P_{eff}$

or $\boxed{S_b = P_{eff} \times f_s}$ (f_s being factor of safety)

Similarly to avoid failure of gear tooth due to wear, wear strength: $S_w > P_{eff}$

or $\boxed{S_w = P_{eff} \times f_s}$

WORKED EXAMPLES

1. For given set of parameters for a pair of helical gears, determine the power transmitting capacity of gears.

Given

T_p = Number of teeth on pinion = 20

T_g = Number of teeth on gear = 100

N_p = Pinion speed = 720 rpm

ψ_n = 20° (normal pressure angle)

α = Helix angle = 25°

b = Face width = 40 mm.

m_n = Normal module = 4 mm

S_{ut} = Ultimate tensile strength = 600 N/mm²

BHN = Hardness number = 300

$C_S = 1.5$, $F_s = 2$

SOLUTION:

Formative no. of teeth (for helical gears)

for opinion $\quad T_{E(P)} = \dfrac{T_p}{\cos^3 \alpha} = \dfrac{20}{\cos^3 (25)} = 26.87.$

So Lewis form factor (based on formative number of teeth from table)

$Y'_{26} = 0.344$

$Y'_{27} = 0.348$

So by interpolation:

$$Y'_{26.87} = 0.344 + (0.348 - 0.344) \times \dfrac{(26.87 - 26)}{(27 - 26)}$$

$\Rightarrow \qquad Y'_{26.87} = 0.3475$

So beam strength $(S_b) = \sigma_b \times b \times m_n \times Y'$

$$= \left(\frac{S_{ut}}{3}\right) \times b \times m_n \times Y'$$

$$= \left(\frac{600}{3}\right) \times 400 \times 4 \times 0.3475 = 11{,}120 \text{ N}$$

and wear strength $= (S_w) = \left(\dfrac{d_p b Q k}{\cos^2 \alpha}\right)$

For $\qquad d_p \rightarrow$ Pinion's pitch circle diameter

since $\qquad m = \dfrac{d_p}{T_p}$ and $m_n = m \cos \alpha$

$\Rightarrow \qquad m_n = \dfrac{d_p}{T_p} \cos \alpha$ or $d_p = \dfrac{m_n T_p}{\cos \alpha}$

$$d_p = \frac{4 \times 20}{\cos(25)} = 88.27 \text{ mm}$$

$\Rightarrow \qquad S_w = (88.27) \times (40) \times \left(\dfrac{2T_g}{T_g + T_p}\right) \times \left(0.16\left(\dfrac{BHN}{100}\right)^2\right)$

$$= (88.27) \times (40) \times \left(\frac{2 \times 100}{100 + 20}\right) \times \left(0.16\left(\frac{300}{100}\right)^2\right)$$

$$= 10{,}318.58 \text{ N}.$$

Since $S_w < S_b \Rightarrow$ (pitting is the criterion of failure.)

So wear strength will be taken as design criteria.

So $S_w = P_{eff} \times (f_s)$

Now $\qquad P_{eff} =$ Effective tooth load

$$= \frac{C_S}{C_V} P_t$$

$$= \frac{1.5}{(5.6/5.6 + \sqrt{v})} \times P_t$$

$$V = \frac{\pi d N}{60} = \frac{\pi \times 88.27 \times 720}{60 \times 10^3} = 3.328 \text{ m/sec}.$$

So $\qquad S_w = \dfrac{1.5}{3.328} \times P_t \times 2 = 10{,}318.58$

$\Rightarrow \qquad \boldsymbol{P_t = 2{,}594.43 \text{ N}.}$

Now as $\qquad M_t = P_t \times \dfrac{d_p}{2} = 2{,}594.43 \times \dfrac{88.27}{2} = 114{,}505.39 \text{ N-mm}$

So $\qquad M_t = \dfrac{60 \times 10^6}{2\pi N_p}(P_{kW}) \Rightarrow 114{,}505.39 = \dfrac{60 \times 10^6 \times P_{kW}}{2\pi \times 720}$

$\Rightarrow \qquad P_{kW} = \dfrac{114{,}505.39 \times 2\pi \times 720}{60 \times 10^6}$

or $\qquad \boxed{P = 8.63 \text{ kW}}$

2. For given set of parameters, design pair of helical gears.

$$T_p = 24, N_p = 5,000 \text{ rpm}, P_{kW} = 2.5 \text{ kW}, V_r \text{ (velocity ratio)} = 4.1$$
$$\psi_n = 20° \text{ (normal pressure angle)}, \alpha = 23°,$$
$$S_{ut} = 750 \text{ N/mm}^2 \text{ (for pinion and gear both)}$$
$$C_S = 1.5, f_s = 2$$

SOLUTION: Assumption:

For preliminary design, considering
Grade 4 gears with face width $(b) = 10 m_n$
and peripheral velocity $(V) = 10$ m/sec.

So transmitted torque $(M_t) = \dfrac{60 \times 10^6 \times (P_{kW})}{2\pi N_p}$

$\Rightarrow$
$$M_t = \frac{60 \times 10^6 \times 2.5}{2\pi \times 5,000} = 4,774.648 \text{ N-mm}$$

also $M_t = P_t \times d_p/2, \quad d_p = mT_p = \dfrac{m_n}{\cos\alpha} T_p$

$\Rightarrow$
$$d_p = \frac{m_n \times 24}{\cos(23)} = \frac{24 \, m_n}{\cos(23)}$$

So
$$M_t = P_t \times \frac{24 m_n / \cos(23)}{2}$$

$\Rightarrow$
$$4,774.648 = \frac{P_t}{2} \times \frac{24 \, m_n}{\cos(23)} \Rightarrow P_t = \frac{366.25}{m_n}$$

So effective load on gear tooth,

$$P_{eff} = \frac{C_S}{C_V} \times P_t$$

Here
$$C_S = 1.5 \text{ and } C_V = \frac{5.6}{5.6 + \sqrt{V}} = \frac{5.6}{5.6 + \sqrt{10}} = 0.6391$$

$\Rightarrow$
$$P_{eff} = \left(\frac{1.5}{0.6391}\right) \times \left(\frac{3,666.25}{m_n}\right)$$

$$= \left(\frac{859.61}{m_n}\right) \text{N}.$$

Now since beam strength $(S_b) = \sigma_b b m_n Y'$

So
$$\left(\frac{859.61}{m_n}\right) \times f_s = T_b b m_n Y'.$$

Now, Y' is based on formative no. of teeth

$$T_{E'_{(P)}} = \frac{T_p}{\cos^3 \alpha} = \frac{24}{\cos^3 (23)} = 30.77$$

by interpolation (from table)

$$Y_{30} = 0.358, \text{ and } Y_{32} = 0.364$$

So
$$Y'_{30.77} = 0.358 + \frac{(0.364 - 0.358)(30.77 - 30)}{(32 - 30)}$$

$\Rightarrow \qquad\qquad Y' = 0.36$

So as $S_b = P_{eff} \times f_s \Rightarrow \left(\dfrac{859.61}{m_n}\right) \times 2 = \left(\dfrac{750}{3}\right) \times 10 m_n \times m_n \times 0.36$

(for effective design) $\Rightarrow m_n = 1.24$ mm.

Taking $\qquad m_n = 1.5$ mm $\Rightarrow P_t = \dfrac{366.25}{m_n} = \dfrac{366.25}{1.5} = 244.16$ N

So $\qquad\qquad b = 10 m_n = 10 \times 1.5 = 15$ mm.

So pitch circle diameter of pinion $(d_p) = \dfrac{m_n T_p}{\cos \alpha} = \dfrac{1.5 \times 24}{\cos(23)}$

$\Rightarrow \qquad\qquad \boxed{d_p = 39.11 \text{ mm}}$

and d_g = Gear's pitch circle diameter

$$= \dfrac{m_n T_g}{\cos \alpha} = \dfrac{1.5 \times 96}{\cos(23)} = \boxed{156.44 \text{ mm}}$$

Here T_g is obtained as:

Velocity ratio $(V_r) = \dfrac{V_p}{V_g} = \dfrac{T_g}{T_p} \Rightarrow 4 = \dfrac{T_g}{24} \Rightarrow T_g = 24 \times 4 = 96$

So $\qquad\qquad \boxed{d_g = 156.44 \text{ mm}}$

Now, finding dynamic load (P_d) and effective load

$$P_{eff} = C_S P_t + P_d$$

also beam strength $= m_n b \sigma_b Y'$

$$\boxed{S_b = 1.5 \times 15 \times 250 \times 0.36 = 2{,}025 \text{ N}}$$

Now $\qquad P_d = \dfrac{21V\left(Ceb\cos^2 \alpha + P_t\right)\cos \alpha}{21V + \sqrt{\left(Ceb\cos^2 \alpha + P_t\right)}}$

where $\qquad e = e_p + e_g$

$\qquad\qquad e_p = 3.20 + 0.25\ \phi_p = 3.20 + 0.25\ (m_n + 0.25\sqrt{d_p})$

$\qquad\qquad\quad = 3.20 + 0.25\ (1.5 + 0.25\sqrt{39.11})$

$\qquad\qquad\quad = 3.9659$ μm

Similarly, $\qquad e_g = 3.20 + 0.25\ \phi_g = 3.20 + 0.25\ (m_n + 0.25\sqrt{d_g})$

$\qquad\qquad\quad = 4.3567$ μm

So $\qquad\qquad e = e_p + e_g = 8.3226$ μm $= 8.3226 \times 10^{-3}$ mm

$\qquad\qquad C = 11{,}400$ N/mm^2

(For 20° full depth involute, pinion and gear made up of steel, from table)

now $\qquad V = \dfrac{\pi d_p N_p}{60 \times 10^3} = \dfrac{\pi \times 39.11 \times 5{,}000}{60 \times 10^3} = 10.24$ m/sec.

So $\qquad P_d = \dfrac{21 \times (10.24) \times (11{,}400 \times 8.3226 \times 10^{-3} \times 15 \cos^2(23) + 244.16) \cos(23)}{21 \times (10.24) + \sqrt{11{,}400 \times 8.3226 \times 10^{-3} \times 15 \cos^2(23) + 244.16}}$

$\Rightarrow$ $P_d = 1{,}133.94$ N

So $P_{eff} = C_S P_t + P_d = 1.5\ (244.16) + 1{,}133.94$

                $= 1{,}500.18$ N

as $S_b = P_{eff} \times f_s \Rightarrow f_s = \dfrac{S_b}{P_{eff}} = \dfrac{2025}{1500.18} = 1.35$

Since factor of safety is normal, design can be considered as safe.

Now, considering wear strength of gear:

             $S_w = P_{eff} \times f_s$, assuming factor of safety $= 2$

So $S_w = 1{,}500.18 \times 2.0 = 3{,}000.36$ N

Now Q = Ratio factor $= \dfrac{2T_g}{T_g + T_p} = \dfrac{2 \times 96}{96 + 24} = 1.6$

and $S_w = \dfrac{bQd_p K}{\cos^2 \alpha} \Rightarrow 3000.36 = 15 \times 1.6 \times 39.11 \dfrac{\left[0.16 \left(\dfrac{BHN}{100} \right)^2 \right]}{\cos^2 (23)}$

$\Rightarrow$ BHN = Hardness number $= 411.44$ or 412

PREVIOUS YEAR UNIVERSITY QUESTIONS

1. A 20° normal pressure angle of helical pinion having 20 teeth and helix angle of 30° transmits 3 kW at 30 rev/sec. The speed ratio is 4, normal module is 4 mm and the face width is 0.36 mm. Calculate the maximum contact stress in the tooth, if the gear and pinion are made up of steel.

                                                    [UPTU 2005]

SOLUTION:

              Pressure angle $(\psi) = 20°$

T_p = Number of teeth on pinion $= 20$

                Helix angle $(\alpha) = 30°$

Power $P = 3$ kW, $N_p = 30$ rev/sec $= 1{,}800$ rpm

$$\text{Velocity ratio} = \frac{d_g}{d_p} = \frac{T_g}{T_p} = \frac{N_p}{N_g} = 4$$

            Module $(m) = 4$ mm and face width $= 36$ mm.

Material of both pinion and gear is same, on this account, pinion will be considered weaker and design will be based upon it.

$$\text{Torque transmitted } (M_t) = \frac{60 \times 10^3 \times (P_{kW})}{2\pi N_p}$$

                       $= 15.92$ N-m

Pitch diameter of pinion $(d_p) = m T_p = 4 \times 20 = 80$ mm $= 0.08$ m

So tangential tooth load (P_t) on pinion

$$= \frac{M_t \times 2}{d_p} = \frac{15.92 \times 2}{0.08} = 398 \text{ N}$$

Formative or equivalent no. of teeth for pinion

$$T_E = \frac{T_p}{\cos^3 \alpha} = \frac{10}{\cos^3 (30)} = 30.79°$$

Assuming teeth to be of 20° stub teeth form, Lewis form factor $(Y_p') = \left(0.175 - \dfrac{0.841}{T_E}\right)$

[From table]

$$= \left(0.175 - \frac{0.841}{30.79}\right) = 0.1477$$

Now peripheral velocity $(V) = \dfrac{\pi d_p N_p}{60} = \dfrac{\pi \times 0.08 \times 1{,}800}{60}$

$$= \frac{\pi \times 0.08 \times 1{,}800}{60} = 7.536 \text{ m/sec.}$$

this is less than 12.5 m/sec.

So velocity factor $(C_V) = \dfrac{4.5}{4.5 + V}$ [assuming teeth are carefully cut]

$\Rightarrow$ $\qquad C_V = \dfrac{4.5}{4.5 + 7536} = 0.374$

So tangential tooth load $\quad (W_t) = (\sigma C_V) \times b\pi\, m(Y_p')$

$\Rightarrow \qquad\qquad\qquad 398 = \sigma \times 0.374 \times 36 \times 3.14 \times 4 \times 0.1477$

$\Rightarrow \qquad\qquad\qquad \sigma = 16 \text{ N/mm}^2$

So contact stress $\qquad\qquad \sigma = 16 \text{ N/mm}^2$

2. With the help of a sketch, explain how an axial thrust is generated in a helical gear.

[UPTU 2005]

SOLUTION:

In order to have more than one pair of teeth in contact, the tooth displacement (advancement) of end of tooth over other end or overlap should be equal to the axial pitch.

$\Rightarrow \qquad\qquad$ Overlap $= P_c = b \tan\alpha$

Normal tooth load (W_N) has two components 'W_t' and 'W_a'.
'W_t' being tangential and 'W_a' being axial.
also $W_a = W_N \sin\alpha = W_t \tan\alpha$ [from figure below]

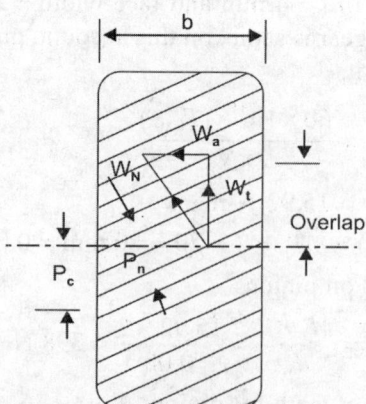

From the above equations, it can be seen that as the helix angle α increases, tooth overlap increases and end thrust given by 'W_a' also increases, which is undesirable.

So overlap should be around 15% of circular pitch for proper design.

So overlap $= b \tan\alpha = 1.15\, P_c \Rightarrow b = \dfrac{1.15\pi m}{\tan\alpha}$

where b = Minimum face width and

$\quad$ m = Module

3. A pair of helical gears are used to transmit 18 kW at 8000 rpm of the pinion. The teeth are 20° stub in diametral plane and the helix angle is 45°. The gear and the pinion have a pitch diameter of 320 mm and 80 mm respectively. Both gear and pinion are made up of cast steel with allowable static strength of 100 MPa, suggest a suitable module and face width for the gear pain and check the strength of the design in wear. Take modulus of elasticity for cast steel as 2×10^5 MPa and $\sigma_{es} = 618$ MPa. $\hfill$ [UPTU 2006]

SOLUTION: Given:

$$\text{Power } (P) \;=\; 18 \text{ kW}$$
$$\text{Speed of pinion } (N_p) \;=\; 8{,}000 \text{ rpm}$$
$$\text{Pressure angle } (\psi) \;=\; 20° \text{ stub tooth}$$

Helix angle $(\alpha) = 45°$, d_g = Pitch circle diameter of gear = 320 mm = 0.32 m

d_p = Pitch circle diameter of pinion = 80 mm = 0.08 m.

Since both pinion and gear are made up of cast steel

($\Rightarrow$ allowable static strength = 100 MPa

$\quad E = 2 \times 10^5$ MPa, σ_{es} = Endurance strength = 618 MPa)

So pinion will be considered as weaker and design will be based upon it.

Torque transmitted by the pinion $(T) = \dfrac{P \times 60}{2\pi N_p}$

$$= \dfrac{18 \times 10^3 \times 60}{2\pi \times 8{,}000} = 21.49 \text{ N-m}$$

Tangential load on pinion

$$F_t = \dfrac{T}{d_p/2} = \dfrac{21.49}{0.08/2} = 537.25 \text{ N}$$

So number of teeth on pinion

$$T_p = \dfrac{d_p}{m} = \dfrac{80}{m}$$

Equivalent (formative) no. of teeth on pinion

$$\left(T_p\right)_E = \dfrac{T_p}{\cos^3\alpha} = \dfrac{80/m}{\cos^3(45)} = \dfrac{226.4}{m}$$

Tooth form factor for pinion for 20° stub tooth system: $y_p = 0.175 - \dfrac{9.95}{T_p}$

$$= 0.175 - \dfrac{9.95}{226.4/m} = (0.175 - 0.0042m)$$

Peripheral velocity $= \dfrac{\pi d_p N_p}{60} = \dfrac{\pi \times 0.08 \times 8{,}000}{60}$

$$= 33.51 \text{ m/sec}$$

So as velocity factor (assuming precision gears) is

$$= \frac{5.55}{5.55 + \sqrt{V}} \text{ (for } V > 20 \text{ m/sec)}$$

$$\Rightarrow \qquad C_V = \frac{5.55}{5.55 + \sqrt{33.51}} = 0.4895$$

Since the maximum face width 'b' for helical gears is in the range $\rightarrow 12.5m$ to $20m$, when m is module, so taking $b = 12.5m$

From Lewis equation: $F_t = \dfrac{y_p \times C_V \pi b m Y'}{C_S}$

assuming $C_S = 1$

$$\Rightarrow \qquad 537.25 = \frac{\pi \times 100 \times 12.5m \times m \times 0.4895(0.175 - 0.0042m)}{1}$$

$$\Rightarrow \qquad 537.25 = 1,922.26m^2 (0.175 - 0.0042m)$$

Solving this equation $m = 1.28$

taking $\boxed{m = 1.5}$, so face width $(b) = 12.5m$

$$= 12.5 \times 1.5$$

$$\boxed{b = 18.75 \text{ mm}}$$

Checking gear for wear:

So velocity ratio $\qquad V_r = \dfrac{d_g}{d_p} = \dfrac{320}{80} = 4$

Ratio factor $\qquad Q = \dfrac{2T_g}{T_g + T_p} = \dfrac{2V_r}{V_r + 1} = \dfrac{2 \times 4}{4 + 1} = 1.6$

ψ_n = Normal pressure angle

as $\tan \psi_n = \tan \psi \cos \alpha$ (for helical gear)

$$= \tan (20) \cos (45) = 0.2573$$

$$\Rightarrow \qquad \psi_n = 14.4°$$

again, since both pinion and gear are made up of cast steel

So $\qquad E_p = E_g = 2 \times 10^5$ MPa

and $\qquad K = \dfrac{(T_{es})^2 \sin \psi_n}{1.4} \left(\dfrac{1}{E_p} + \dfrac{1}{E_g} \right)$

$$= \frac{(618)^2 \sin (14.4)}{1.4} \left(\frac{1}{2 \times 10^5} + \frac{1}{2 \times 10^5} \right) = 0.678 \text{ N/mm}^2$$

So maximum limiting wear load

$$F_w = \frac{d_p b Q K}{\cos^2 \alpha} = \frac{80 \times 18.75 \times 1.6 \times 0.678}{\cos^2 (45)} = 3,254.4 \text{ N}$$

Since $F_w \gg F_t \Rightarrow$ design in safe from wear consideration.

3. Explain briefly, "formative/equivalent" number of teeth in helical gears.

SOLUTION:

Formative number of teeth for helical gear may be defined as number of teeth generated on the surface of a cylinder having a radius equal to the radius of curvature at a point at tip of minor axis of an ellipse obtained by taking a section of gear in the normal plane.

Mathematically, equivalent no. of teeth on a helical gear $(T_e) = T/\cos^3 \alpha$
where T = Actual number of teeth on helical gear.

T_e = Formative no. of teeth

α = Helix angle

2.5 FORCE ANALYSIS ON HELICAL GEAR TOOTH

Force acting on helical gear tooth can be resolved into three components.
- Tangential component (P_t)
- Radial component (P_n)
- Thrust component (P_a)

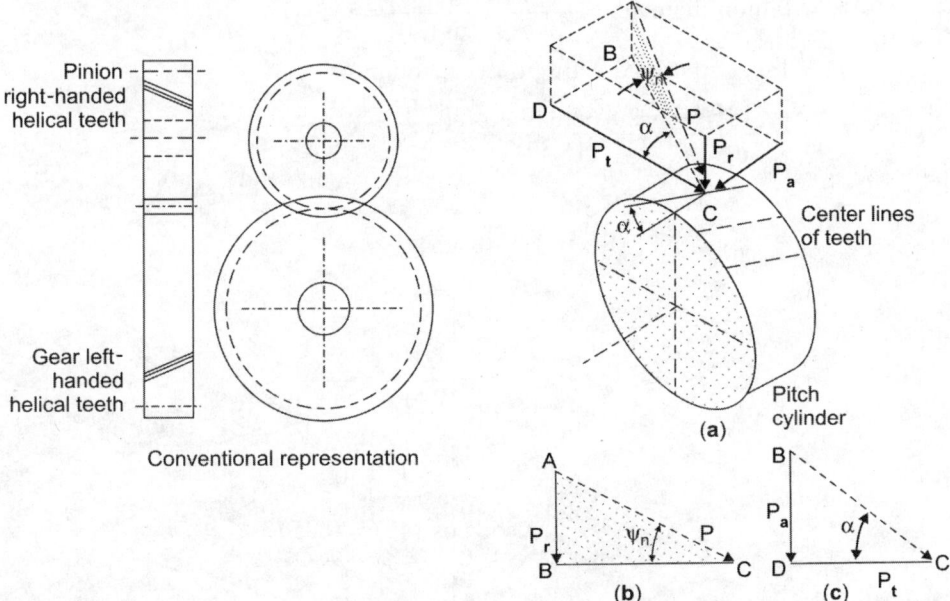

Fig. 2.5 Components of tooth force

  i. Tangential component 'P_t' (as seen in the figure) for pinion or driving gear acts in a direction opposite to the direction of rotation and along tangent to the pitch circle, whereas it acts along the same direction as that of driven gear rotation.
 ii. Radial component 'P_n' acts towards center of pinion and driven gear.
iii. Thrust component 'P_a' acts in a direction as right hand rule for helix with right hand curvature and left hand rule for helix with left hand curvature. Thumb direction will represent axial thrust's direction taken in the same sense as radial and tangential loads on components.

From the figure, it can be proved that

$$\therefore \qquad P_r = P_t \left(\frac{\tan \psi_n}{\cos \alpha} \right)$$

and
$$P_a = P_t \tan \alpha$$

WORKED EXAMPLE

1. For given set of helical gear parameters, determine components of tooth force and show these forces acting on pinion and gear.

P_{kW} = Power transmitted = 5 kW, speed of pinion (N_p) = 720 rpm, m_n = Normal module = 5 mm, ψ_n = Normal pressure angle = 20°, pinion has RH (right-handed) teeth and gear has LH (left-handed) teeth, α = Helix angle = 30°, sense of rotation of pinion is clockwise, T_p = 20, T_g = 30.

SOLUTION:

Torque transmitted through pinion (A)

$$(M_t)_p = \frac{60 \times 10^6 \times P_{kW}}{2\pi N_p} = \frac{60 \times 10^6 \times 5}{2\pi (720)}$$

$$= 66{,}314.56 \text{ N-mm.}$$

$$d_p = \text{Pinion diameter} = \frac{T_p m_n}{\cos\alpha} = \frac{20 \times 5}{\cos(30)} = 115.47\,\text{mm}$$

P_t = Tangential load acting on gear tooth

$$= \frac{2 \times (M_t)_P}{d_p} = \frac{2 \times 66{,}314.56}{115.47} = 1{,}148.6 \text{ N}$$

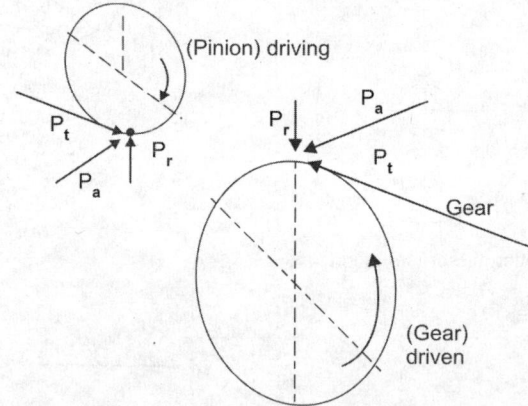

Direction of P_t, P_r and P_a on pinion and gear respectively

axial thrust (P_a) = $P_t \tan\alpha$ = 1,148.6 tan(30) = 663.14 N.

and radial load $(P_r) = P_t \left(\dfrac{\tan\psi_n}{\cos\alpha} \right) = 1{,}148.6 \times \dfrac{\tan(20)}{\cos(30)}$

$$= 482.73 \text{ N.}$$

2.6 HERRINGBONE GEARS

Herringbone gears are constructed by joining two identical gears, having teeth with the opposite hand of helix, but same module, number of teeth and pitch circle diameter.

This construction eliminates the axial thrust component of the incident load on gear tooth in case of helical gears which impose reactions on the bearings.

These gears are cut by two cutters which reciprocate (to and fro motion) along a straight line out of phase to avoid clashing.

Net axial force acting on bearings is zero and the power transmitting capacity is high, along with high peripheral speeds attainability.

These gears find application in high power applications like ship drives and turbines. Helix angle varies between 20° and 45° and requires high precision for locating Herringbone gears axially on shaft, for equal distribution of load between right and left parts of gear. So as to balance thrust forces as well as high cost can be considered as a limitation of the gear.

Design criteria is same as helical gears with the consideration that Herringbone gear is equivalent to two identical helical gears, transmitting one half power each.

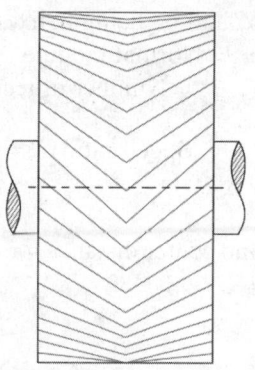

Fig. 2.6 Herringbone gear

2.7 CROSSED HELICAL GEARS

For connecting two shafts whose axes are neither parallel nor intersecting, helical gears used are called crossed helical gears.

The fundamental difference of crossed helical gears from (parallel or normal) helical gears is there being 'point contact' between meshing teeth of crossed helical gears in contrast to 'line contact' in case of parallel helical gears.

These gears have low load carrying capacity on this account (point contact), as contact area being very small, contact pressure and subsequently wear are high.

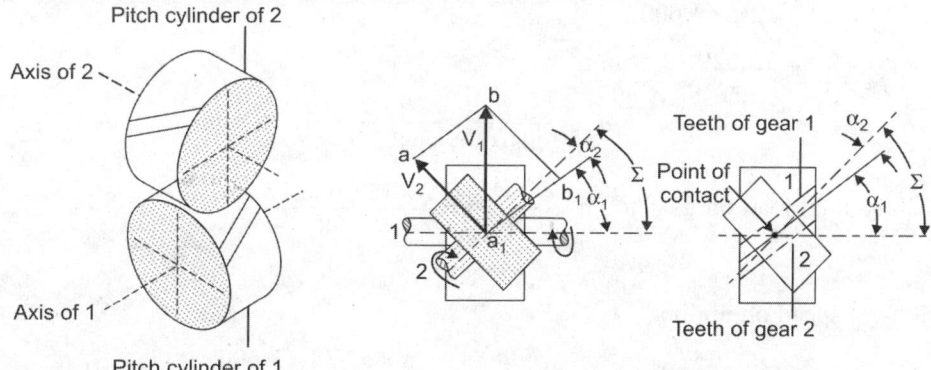

Fig. 2.7 Crossed helical gears   **Fig. 2.8** Crossed helical gear—sliding velocity

So these gears find application in light duty applications like small IC engines, oil pump and distribution systems, driving speedometer cable, feed mechanisms of machine tools, etc.

There is no difference in construction between crossed helical gear and parallel helical gear, until they are mounted in a particular position.

At the point of contact, a sliding velocity vector is generated which is the vector difference between peripheral velocities of the 2 gears present in crossed helical gears.

For crossed helical gears:

$$d_p = \frac{T_p m_n}{\cos \alpha_p} \text{ and } d_g = \frac{T_g m_n}{\cos \alpha_g}$$

α_p and α_g being helix angles of pinion and gear respectively.

m_n = Normal module

T_p, T_g = Number of teeth on pinion and gear respectively.

$$\text{Speed ratio} \ (i) = \frac{N_p}{N_g} = \frac{T_g}{T_p} = \frac{d_g \cos \alpha_g}{d_p \cos \alpha_p}$$

and center distance $= a = \dfrac{d_p}{2} + \dfrac{d_g}{2} = \dfrac{1}{2} \times m_n \left[\dfrac{T_p}{\cos \alpha_p} + \dfrac{T_g}{\cos \alpha_g} \right]$ between crossed helical gears.

WORKED EXAMPLES

1. For a Herringbone speed reducer having a normal module of 2 mm and $T_p = 26$, $T_g = 104$, pressure angle (ψ) = 20°, helix angle (α) = 25°, pinion receives 100 kW power and rotation (N_p) at 3,600 rpm.

Face width of each half being 35 mm. Material being alloy steel (30 Ni4Cr1) with ultimate tensile strength being S_{ut} = 1,500 N/mm², surface hardness = 450 BHN, C_S = 1.25.

Determine factor of safety against bending and pitting failures.

SOLUTION:

Torque transmitted $(M_t) = \dfrac{60 \times 10^6 \times P_{kW}}{2 \pi N_p}$

$\qquad = \dfrac{60 \times 10^6 \times 50}{2 \pi \times 3,600} = 0.13262.12 \,\text{N-mm (half of 100 kW as per design criteria)}$

$\qquad = 13,629.12 \,\text{N-mm}.$

So $\qquad\qquad d_p = \dfrac{T_p m_n}{\cos \alpha_p} = \dfrac{26 \times 2}{\cos (25)} = 57.38 \,\text{mm}$

P_t = Tangential load on gear tooth $= \dfrac{2 \times M_t}{d_p} = \dfrac{2 \times 132,620.12}{57.38}$

$\qquad\qquad = 4,622.83 \,\text{N}$

Peripheral speed on pinion:

$$V = \frac{\pi d_p N_p}{60 \times 10^3} = \frac{\pi \times 57.38 \times 3,600}{60 \times 10^3} = 10.82 \,\text{m/sec}$$

So coefficient of velocity $\quad (C_V) = \dfrac{5.6}{5.6 + \sqrt{V}} = \dfrac{5.6}{5.6 + \sqrt{10.82}} = 0.63$

So $\qquad\qquad P_{eff} = \dfrac{C_S \times P_t}{C_V} = \dfrac{1.25 \times 4,622.83}{0.63} = 9,172.28 \,\text{N}$

Now formative no. of teeth for driving gear (pinion):

$$T_{E_p'} = \frac{T_p}{\cos^3 \alpha} = \frac{26}{\cos^3 25} = 34.93$$

and for $T_{E_p'}$ = 34.93, Lewis form factor Y_p' = 0.37279 (from interpretation)

So beam strength $(S_b) = \sigma_b m_n b Y_p'$

$$= \left(\frac{S_{ut}}{3}\right) m_n b Y_p'$$

$$= \left(\frac{1,500}{3}\right) \times 2 \times 35 \times 0.37279$$

$$= 13,047.65 \text{ N}$$

and wear strength $= \dfrac{bQd_p K}{\cos^2 \alpha}$

$$Q = \frac{2T_g}{T_g + T_p} = \frac{2 \times 104}{104 + 26} = 1.6$$

BHN = 450, $\quad K = 0.16\left(\dfrac{BHN}{100}\right)^2 = 0.16\left(\dfrac{450}{100}\right)^2 = 3.24 \,\text{N/mm}^2.$

So wear strength $\quad (S_w) = \dfrac{bQd_p K}{\cos^2 \alpha}$

$$= \frac{35 \times 1.6 \times 57.38 \times 3.24}{\cos^2 (25)}$$

$$= 12,674.83 \text{ N}$$

So factor of safety:

$$f_s = \frac{S_b}{P_{eff}} = \frac{13,047.65}{9,172.28} = 1.42 \text{ (considering beam strength a bending failure)}$$

and $f_s = \dfrac{S_w}{P_{eff}} = \dfrac{12,674.83}{9,172.28} = 1.38$ (considering pitting failure or wear strength)

2. Explain briefly Herringbone gear's advantage over parallel helical gears.

SOLUTION:
The main disadvantage of single helical gears is thrust on the shaft bearings. The side thrust on bearings can be eliminated by using the helical gears in pairs, one having the right hand helix and other having the left hand helix. These gears are mounted on shaft side by side and the axial thrust produced by one gear is neutralized by the other. If both gears are formed on one blank, the resulting gear is called Herringbone gear.

3

Worm Gears

3.1 INTRODUCTION

Worm gears are used to transmit power between two non-intersecting shafts which are at right angles to each other. Worm gear drive consists of two mating parts having a line of contact between them. These parts are named: 'Worm'—A toothed screw consisting of single/multistart threads and 'Worm wheel'—Toothed gears. As the worm (driver) rotates, its threads exert pressure on teeth of the worm gear (driven) to impart its rotational motion. Worm gear drives offer following advantages over other gear drives.

- High speed reduction up to 100:1
- Operation being smooth and silent
- Drive (worm gear) is compact as compared to other gear drive
- Self locking as in case of cranes and lifting devices can be achieved, where motion is transmitted from the worm to the wheel.

Some of the shortcomings (disadvantages) with worm gear drive being:

- Efficiency of power transmission is quite low.
- Used for low power transmission up to 100 kW.
- Use of materials like phosphor bronze in worm gear drive make it quite expensive.
- High generation of heat during operation requires heat to be dissipated to lubricate oil and surroundings.

Suitability of worm gear drive is decided on the basis of requirements of mechanical advantages and efficiency.

a. High mechanical advantage, but low efficiency requirements—single-threaded worm.
b. Low mechanical advantage, but high efficiency requirements—multithreaded worm.

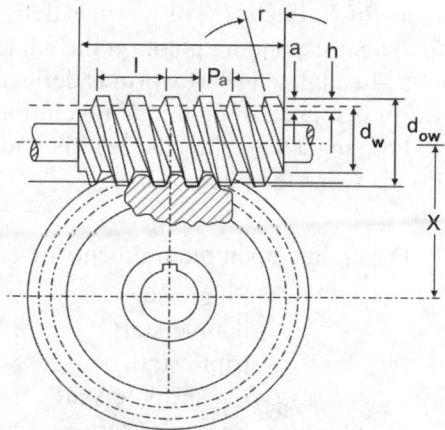

Single-threaded worms find application in manually operated intermittent mechanism, steering mechanism and operating (opening and closing) gate valves, motorized opening and closing of small hoists and large gate valves, etc.

Whereas multithreaded worms find application in motorized continuous operation drives for machine tools, motorized speed increasing applications and automotive supercharge and centrifugal charger.

Fig. 3.1 Worm and worm gear

3.2 TYPES OF WORM

a. Single enveloping worm: 'Worm gear' partially encloses the 'worm' and have a line of contact between worm and worm gear. Also called cylindrical worm and is more widely used.

b. Double enveloping worm: 'Worm' as well as 'worm gear' enclose each other and have an area of contact (threads of worm and teeth of worm wheel) among them. Also called as 'hour glass' worm and drive as 'cone gearing'

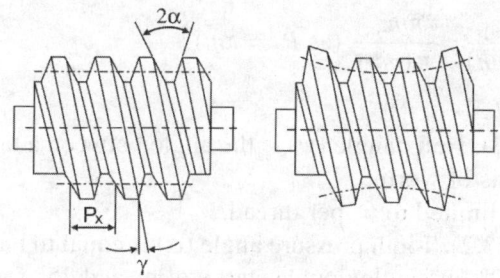

Fig. 3.2 Single and double enveloping worm

Double enveloping worm gear drive has a low contact pressure between the threads and teeth, thereby reducing wear. Also this drive requires two-thirds of the space and one-third of the weight compared with single enveloping worm gear drive.

However, this drive (double enveloping) requires precise alignment. Small deviation from center results in loss of theoretical area of contact.

3.3 TERMINOLOGY

Worm gear drives are designated by four quantities in the following manner:

$$z_1/z_2/q/m$$

where
z_1 = Number of starts on the worm
z_2 = Number of teeth on worm wheel
q = d_1/m (diametral quotient)
m = Module (mm)

(**Note:** Number of starts on the worm (z_1) refers to single-threaded or multi-threaded worm. While single-threaded worm gives large speed reduction, but efficiency is low, whereas multi-threaded worm gives high efficiency, but speed reduction is low.)

a. Pitch: Pitch (P_x) of worm is defined as the distance measured from a point on one thread to the corresponding point on the adjacent thread along the axis of worm.

b. Lead: Lead (l) of worm is defined as the distance that a point on thread will move, when worm is rotated through one revolution.

For single start threads, lead is equal to the axial pitch, whereas for double start threads, lead is twice the pitch.

So
$$l = P_x \, z_1$$

Depending upon requirements of velocity ratio, single- or multi-threaded worm is used.

Single start	:	Velocity ratio 20 and above
double start	:	12–36
triple start	:	8–12
Quadrupole start	:	6–12
Sextuple start	:	4–10

(**Note:** Single-threaded worm refers to a thread length comprising of a single thread in one revolution whereas a multithread say triple start/threaded worm refers to a thread length comprising of three threads in one revolution.)

c. Lead angle: Lead angle (γ) is the angle between tangent to the thread at the pitch diameter and a plane normal to worm axis such that:

From Fig. 3.3.1,
$$\tan \gamma = l/\pi \, d_1$$

$\Rightarrow \qquad \tan \gamma = \dfrac{P_x z_1}{\pi(qm)} = \dfrac{\pi m z_1}{\pi(qm)} \text{ (as } P_x = \pi m)$

$\Rightarrow \qquad \tan \gamma = z_1/q$

d. Helix angle (Fig. 3.3): Helix angle (ψ) is the angle between a tangent to the thread at the pitch diameter and the axis of worm.

Helix angle should be limited to 6° per thread.

e. Pressure angle (Fig. 3.2): Tooth pressure angle (α) is equal to half of the thread angle and should not be less than 20° for single/double start worms and 25° for triple/multistart worms.

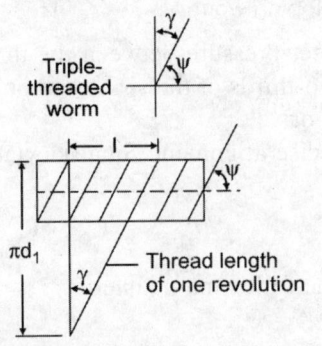

Fig. 3.3 Triple-threaded worm

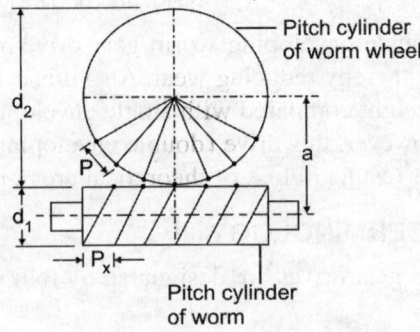

Fig. 3.4 Pitch cylinder (worm)

f. Center distance (a): Center distance between worm and worm wheel is given by:

$$a = \frac{1}{2} m (q + z_2)$$

q: Diametral quotient $= d_1/m$

z_2: Number of teeth on worm wheel

g. Speed ratio (i): It is the ratio of no. of teeth on worm wheel to the no. of starts on worm.

$$\text{i.e. } i = z_2/z_1$$

On one complete revolution of worm wheel, worm completes z_2 (number of worm wheel teeth) revolution for single start whereas for double start threads, no. of revolution will be ($z_2 / 2$).

3.4 GEAR TOOTH PROPORTIONS

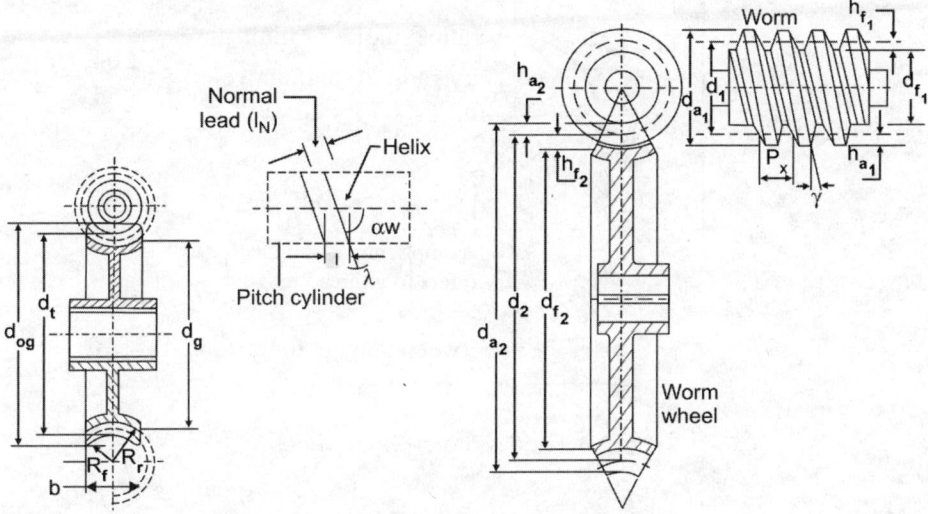

Fig. 3.5 Worm gear dimensions

For an involute helicoidal tooth form:

$$\text{Addendum } (h_{a_1}) = m$$
$$\text{Dedendum } (h_{f_1}) = (2.2 \cos \gamma - 1)\, m$$
$$\text{Clearance } (C) = 0.2\, m \cos \gamma$$

So addendum and dedendum diameters (of worm) are expressed as:

$$\text{addendum diameter } (d_{a_1}) = d_1 + 2h_{a_1} = qm + 2m$$

or $d_{a_1} = m\,(q + 2)$

Dedendum diameter $\left(d_{f_1}\right) = d_1 - 2h_{f_1} = qm - 2m\,(2.2 \cos \gamma - 1)$

$\Rightarrow\ d_{f_1} = m\,(q + 2 - 4.4 \cos \gamma)$

Now for worm wheel, the dimensions can be similarly expressed as:

$$\text{Addendum } \left(h_{a_1}\right) = m\,(2 \cos \gamma - 1)$$
$$\text{Dedendum } \left(h_{f_2}\right) = m\,(1 + 0.2 \cos \gamma)$$

So addendum diameter

$$d_{a_2} = d_2 + 2h_{a_2} = m\,(z_2 + 4 \cos \gamma - 2)$$

And dedendum diameter

$$d_{f_2} = d_2 - 2h_{f_2} = m\,(z_2 - 2 - 0.4 \cos \gamma)$$

Effective face width

$$F = 2m\sqrt{1 + q}$$

[m = Module, q = Diametral quotient (d_1/m)]

Length of root of worm wheel teeth

$$l_r = \left(\frac{2\delta}{2\pi}\right)\left[\pi\left(d_{a_1} + 2C\right)\right]$$ (from Fig. 3.6)

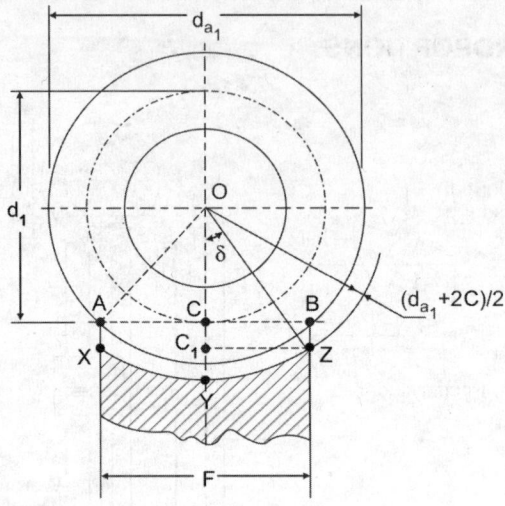

Fig. 3.6 Face width of worm wheel

and
$$\delta = \sin^{-1}\left(\frac{F}{d_{a_1} + 2C}\right)$$

So
$$l_r = \left(d_{a_1} + 2C\right)\sin^{-1}\left[\frac{F}{\left(d_{a_1} + 2C\right)}\right]$$

3.5 EFFICIENCY OF WORM GEARS

Efficiency (η) of worm gears is the ratio of work done by the worm gear to the work done by the worm.

$$\eta = \frac{\tan\lambda\left(\cos\phi - u\tan\lambda\right)}{\cos\phi\tan\lambda + \mu}$$

where
μ = Coefficient of friction
ϕ = Normal pressure angle
λ = Lead angle

For maximum efficiency: $\tan\lambda = \left(\sqrt{1 + \mu^2} - \mu\right)$

Coefficient of friction $(\mu) = \dfrac{0.275}{(V_r)0.25}$ for contact/peripheral speed $(V_r) = \dfrac{\pi d_w N_w}{\cos\lambda}$ between 12 and 180 m/sec

and $\mu = 0.025 + \dfrac{V_r}{18,000}$ for $V_r > 180$ m/min.

Efficiency of worm gears is low (50–98%) and is inversely proportional to speed ratio, provided 'μ' is constant.

If $\mu > \tan\lambda$

$\Rightarrow$ Self-locking drive

i.e. 'worm' is 'driver' and 'worm wheel' is 'driven' and reverse motion is not possible, if $\mu < \tan \lambda \Rightarrow$ reversible drive.

3.6 HEAT DISSIPATION IN WORM GEARING

In worm gearing, the work done due to friction (between worm and worm wheel) is converted into heat. On continuous operation, considerable amount of heat is generated.

* Rate of heat generated is given by:

$$\boxed{H_g = 1,000\,(1-\eta)\,P_{kW}}$$

where 'H_g' is the rate of heat generated.

'η' = Efficiency of worm gears

and $\quad P_{kW}$ = Power transmitted by gears (in kW)

This heat 'H_g' is dissipated through the lubricating oil to the housing wall and finally to surrounding air.

* Rate of heat dissipation:

$$\boxed{H_d = K(t - t_0)A}$$

where $\quad K$ = Overall heat transfer coefficient of housing walls (W/m^2 °C) [12–18 W/m^2 °C]

t = Temperature of lubricating oil (°C)

t_0 = Temperature of surrounding air (°C)

A = Effective surface area of housing (m^2)

So for balance of heat:

$H_g = H_d$

$\Rightarrow \qquad 1,000\,(1-\eta)\,P_{kW} = K(t-t_0)\,A \Rightarrow \boxed{P_{kW} = \dfrac{K(t-t_0)A}{1,000\,(1-\eta)}}$

$\Rightarrow \qquad \boxed{t = t_0 + \dfrac{1,000\,(1-\eta)\,P_{kW}}{KA}}$

P_{kW} is the power transmitting capacity based on thermal considerations.

WORKED EXAMPLES

1. Calculate center distance, speed reduction, worm and worm wheel dimensions for worm gear designated as: 1/30/1018

SOLUTION:

$\Rightarrow \quad z_1 = 1, z_2 = 30, q = 10, m = 8$

$\Rightarrow \quad \dfrac{d_1}{m} = 10 \Rightarrow d_1 = 10 \times m = 10 \times 8 = 80$ mm

Center distance $(a) = \dfrac{1}{2}m(q + z_2)$

$= \dfrac{1}{2} \times 8\,(10 + 30) = 160\,\text{mm}$

Speed ratio $\quad (i) = z_2/z_1 = \dfrac{30}{1} = 30$

Worm dimensions: d_1 (PCD) = 80 mm

Addendum diameter $= d_{a_1} = m(q + 2) = 8\,(10 + 2) = 96$ mm

$$\tan \gamma = z_1/q = \frac{1}{10} \text{ and } \gamma = 5.71° \text{ (lead angle)}$$

Dedendum diameter

$$d_{f_1} = m(q + 2 - 4.4 \cos\gamma) = 8(10 + 2 - 4.4 \cos(5.71°))$$
$$= 60.9747$$
$$P_x = \pi m = \pi (8) = 25.13 \text{ mm}$$

Worm wheel dimensions:

Pitch circle diameter (PCD) $(d_2) = d_{f_2} = 8 (30) = 240$ mm

Addendum diameter $\left(d_{a_2}\right) = m(z_2 + 4 \cos \gamma - 2) = 8(30 + 4 \cos (5.71° - 2))$
$$= 255.8412 \text{ mm}$$

Dedendum diameter $\left(d_{f_2}\right) = m(z_2 - 2 - 0.4 \cos \gamma)$
$$= 8 (30 - 2 - 0.4 \cos (5.71°)) = 220.8159 \text{ mm}.$$

2. Design a high efficiency worm gear speed reducer to transmit continuously the rated power output of 15 kW motor turning at 1,750 rpm. The steel worm with BHN 250 is to be integral with the motor shaft and velocity ratio is 10. The phosphor bronze gear should not have less than 40 teeth. Determine pitch, face and diameter of gears.

SOLUTION: Power $(P) = 15$ kW $= 15 \times 10^3$ watts

Velocity ratio $(V_r) = 10$

$N_w = 11,750$ rpm

Number of teeth on gear >40.

Since the center distance between shafts is not known, therefore, let us assume that for this size unit the center distance $(x) = 100$ mm.

We know that pitch circle diameter of the worm

$$d_w = \frac{(x)^{0.875}}{1.416} = \frac{(100)^{0.875}}{1.416} = 39.71$$

or $d_w = 40$ mm

Pitch circle diameter of the worm gear

$$d_g = 2x - d_w = 2 \times 100 - 40 = 160 \text{ mm}$$

(From Design Data Handbook, for a transmission ratio of 10, quadruple or sextuple threads pointing to number of starts in worm as 4 and 6 respectively, may be used.)

Considering quadruple thread

Number of teeth on worm $(T_g) = 4 \times 10 = 40$

Also axial pitch of the thread on worm gear (P_a)

$$P_a = P_c = \frac{\pi d_g}{T_g} = \frac{\pi \times 160}{40} = 12.57 \text{ mm}$$

Module $(m) = \frac{P_c}{\pi} = \frac{12.57}{\pi} = 2.963$ say 3 mm

Actual circular pitch $(P_c) = \pi m = \pi \times 3 = 9.425$ mm

Actual pitch circle diameter of the worm gear

$$d_g = \frac{P_c \times T_g}{\pi} = \frac{9.425 \times 40}{\pi} = 120 \text{ mm}$$

Actual pitch circle diameter of the worm
$$d_w = 2x - d_g = 2 \times 100 - 120 = 80 \text{ m}$$
Face width of the worm gear (*b*) may be taken as
$$0.73 \times P_{cd} \text{ (worm)}$$
So $\qquad b = 0.73 \times d_w = 0.73 \times 80$
or $\qquad b = 5.84$ mm.

PREVIOUS YEAR QUESTION PAPERS/PROBLEMS (UPTU)

1. Design 20° involute worm and gear which will transmit 15 kW between shafts that are 0.30 m apart, if the speed reduction is to be 10.5 to 1 and the driving shaft is turning at 1,200 rpm.

(UPTU 2006)

SOLUTION: $\qquad$ Given: $\phi = 20°$, $P = 15$ kW $= 15 \times 10^3$ watt
$\qquad$ Worm speed $(N_w) = 1,200$ rpm
$\qquad$ Velocity ratio $(V_r) = 10.5$
$\qquad$ $x = 0.3$ m $= 300$ mm

i. Design of worm
$\qquad$ Let l_N = Normal lead
$\qquad\qquad$ λ = Lead angle
We know that value of x/l_N will be minimum corresponding to
$\qquad\qquad$ $\cot^3 \lambda = V_r = 10.5$.
or $\qquad\qquad$ $\cot \lambda = (10.5)^{1/3} = 2.18$
or $\qquad\qquad$ $\tan \lambda = 0.458 \Rightarrow \lambda = 24.64°$

We know that $\qquad \dfrac{x}{l_N} = \dfrac{1}{2\pi} \left(\dfrac{1}{\sin \lambda} + \dfrac{V_r}{\cos \lambda} \right)$

or $\qquad \dfrac{300}{l_N} = \dfrac{1}{2\pi} \left[\dfrac{1}{\sin (24.64)} + \dfrac{10.5}{\cos (24.64)} \right]$

$\qquad\qquad$ $l_N = 135$ mm

and axial lead $\qquad l = \dfrac{l_N}{\cos \lambda} = \dfrac{135}{\cos (24.64)}$

$\qquad\qquad\qquad$ $= 148.6$ mm

From Data Book, we find that for a velocity ratio of 10.5, the number of starts in threads on the worm $(n) = T_w = 4$

Axial pitch of the threads on the worm
$$P_a = \frac{l}{4} = \frac{148.6}{4} = 37.15 \text{ mm}$$

$\therefore$ module $(m) = \dfrac{P_a}{\pi} = \dfrac{37.15}{3.14} = 11.8$ mm

Let us take the standard value of module $(m) = 8$ mm

Axial pitch of the threads in the worm
$$P_a = \pi m = \pi \times 12 = 37.71 \text{ m}$$

Axial lead of the threads of the worm
$$l = P_a \times n = 37.71 \times 4$$
$$= 150.8 \text{ mm}$$

and normal lead of the threads on the worm (l_N) = Load = 150.8 cos (24.64)

$$= 137 \text{ mm}$$

Center distance $(x) = \dfrac{l_N}{2\pi}\left[\dfrac{1}{\sin\lambda} + \dfrac{V_r}{\cos\lambda}\right]$

$$= \dfrac{1}{2\times 3.14}\left[\dfrac{1}{\sin(24.64)} + \dfrac{10.5}{\cos(24.64)}\right]$$

or $\qquad\qquad x = 304.3$ mm

Let d_w = Pitch circle diameter of the worm

We know that, $\tan\lambda = \dfrac{l}{\pi d_w}$

$\Rightarrow \qquad\qquad d_w = \dfrac{l}{\pi\tan\lambda} = \dfrac{150.8}{3.14\times\tan(24.64)}$

or $\qquad\qquad d_w = 104.7$ mm

Since the velocity ratio is 12 and the worm has quadruple threads $(n = T_w = 4)$, therefore, number of teeth on the worm area

$$T_g = 10.5 \times 4 = 42$$

For 20° pressure angle 4 quadruple threads from Design Data Handbook

Face length of the worm (L_w) $= P_a (4.5 + 0.02\, T_w)$

$$= 37.15 (4.5 + 0.02 \times 4)$$

$$L_w = 170.15 \text{ mm}$$

This length should be increased by 25 to 30 mm for the feed marks produced by the vibrating grinding wheel as it meshes with thread root.

So let $\qquad\qquad L_w = 195$ mm

Depth of teeth

$$h = 0.623\, P_c$$

or $\qquad\qquad h = 0.623\, P_a$

or $\qquad\qquad h = 0.623 \times 37.15 = 23.14$ mm

addendum $\qquad\qquad a = 0.296\, P_c = 0.296 \times 37.15 = 10.6$ m

Outside diameters of worm

$$d_{ow} = d_w + 2a = (104.7 + 2 \times 10.6)$$

or $\qquad\qquad d_{ow} = 126$ mm

ii. Design of worm gear:

Pitch circle diameter of the worm gear

$$d_g = m \times T_g = 12 \times 42 = 504 \text{ m}$$

or $\qquad\qquad d_g = 0.504$ mm

Outside diameter of worm gear:

$$d_{og} = d_g + 0.8903\, P_c$$

$$d_{og} = (504 + 0.8903 \times 37.15)$$

$$d_{og} = 537 \text{ mm}$$

Throat diameter: $\qquad d_T = d_g + 0.572\, P_c = (504 + 0.572 \times 37.15)$

or $\qquad\qquad d_T = 525.25$

and face width $\qquad b = 2.15\, P_c + 5$

or $\qquad\qquad b = (2.15 \times 37.15 + 5)$

or $\qquad\qquad b = 84.8$ mm

2. A worm gear box with an effective surface area of 1.5 m^2 is operating in still air with a heat transfer coefficient of 15 W/m^2 °C. The temperature line of the lubricating oil above the atmospheric temperature is limited to $50°$ C. The worm gears are designated as $1/30/10/8$.

[UPTU 2007] (Special paper)

The worm shaft is rotating at $1,440$ rpm and the normal pressure angle is $20°$. Calculate the power transmitting capacity based on the thermal consideration.

SOLUTION:

Given: $z_1 = 1$, $z_2 = 30$, $q = 10$ and $m = 8$ mm

$$\tan\gamma = \frac{z_1}{q} = \frac{1}{10} = 0.1 \text{ or } \gamma = 5.71°$$

$$d_1 = mq = 8 \times 10 = 80 \text{ m}$$

$$V_S = \frac{\pi d_1 n_1}{60,000 \times \cos\gamma} = \frac{9 \times 80 \times 1,440}{60,000 \times \cos(5.71)} = 6.06 \text{ m/sec}$$

From Data book, $\mu = 0.024$ (for $V_S = 6.06$ m/sec)

so

$$\eta = \frac{(\cos\alpha - \mu\tan\gamma)}{(\cos\alpha + \mu\cot\gamma)}$$

$$= \frac{\cos(20) - 0.024(0.1)}{\cos(20) + 0.024(1/0.1)} = 0.7945$$

So

$$P_{kW} = \frac{K(t - t_0)A}{1,000(1 - \eta)} = \frac{15 \times 50 \times 1.5}{1,000(1 - 0.7945)} = 5.47.$$

3.7 STRENGTH AND WEAR TOOTH LOAD FOR WORM GEARS

Design of worm gear drive on the basis of strength (beam strength) can be based on 'Lewis equation' as applied to teeth of worm wheel.

Maximum permissible torque that the worm wheel withstands without bending, is lower of the values given by:

$$(M_t)_1 = 17.65 X_{b_1} S_{b_1} mlr\, d_2 \cos\gamma$$

$$(M_t)_2 = 17.65 X_{b_2} S_{b_2} mlr\, d_2 \cos\gamma$$

[where (subscript 1) $\Rightarrow$ worm and (subscript 2) $\Rightarrow$ worm wheel]

M_t = Permissible stress
X_b = Speed factor
S_b = Bending stress factor
m = Module
l_r = Length of worm wheel teeth's root
d_2 = Pitch circle diameter of worm wheel
γ = Lead angle of worm

'M_t' is obtained from $P_{kW} = \dfrac{2\pi n M_t}{60 \times 10^6}$

and 'M_t' is the lower value of $(M_t)_1$ and $(M_t)_2$.

Design of worm gear drive on the basis of wear rating (pitting failure) is in terms of maximum permissible torque (lower of the values given below) that worm wheel (weaker of worm and worm wheel) can withstand.

$$(M_t)_3 = 18.64 X_{c_1} S_{c_1} Y_z (d_z) 1.8 \text{ m}$$

$$(M_t)_4 = 18.64 X_{c_2} S_{c_2} Y_z (d_2) 1.8 \text{ m}$$

where M_t is the permissible torque on the worm wheel.

X_{c_1}, X_{c_2} = Speed factors for worm and worm wheel

(depend on peripheral/rubbing speed *vs* rpm)

S_{c_1}, S_{c_2} = Surface stress factors of worm and worm wheel

Y_z = Zone factor

3.8 DESIGN OF WORM GEARING

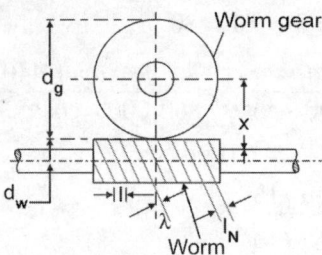

Fig. 3.7 Worm and worm gearing design

Worm gearing design comprising number of threads, lead, lead angle are computed from given quantities like power transmitted, speed, velocity ratio and center distance between shafts.

Center distance $(x) = \dfrac{d_w + d_g}{2}$

By geometry: $x = \dfrac{1}{2}\pi(\cot\lambda + V_r)$

Normal lead (l_N) = $l \cos\lambda$ (l = Lead, λ = Lead angle, V_r = Velocity ratio)

$\Rightarrow \qquad x = \dfrac{l_N}{2\pi}\left(\dfrac{1}{\sin\lambda} + \dfrac{V_r}{\cos\lambda}\right)$

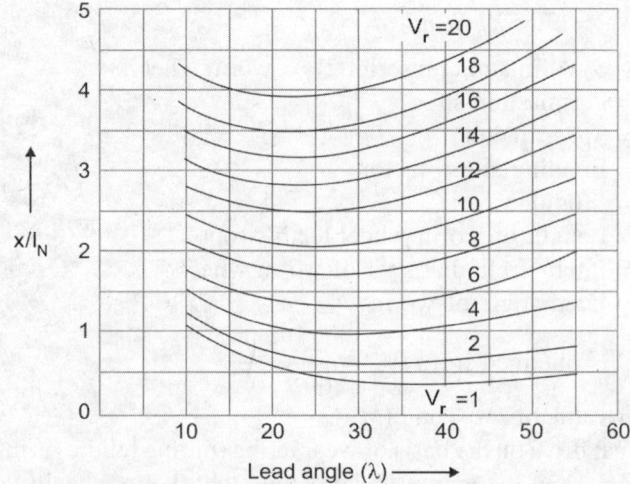

Fig. 3.8 Worm gear design curves

$$\Rightarrow \qquad \frac{x}{l_N} = \frac{1}{2\pi}\left(\frac{1}{\sin\lambda} + \frac{V_r}{\cos\lambda}\right) \qquad\qquad ...(A)$$

Graph shows optimum lead angle for any point on each curve of velocity ratio and corresponds to minimum value of x/l_N. This gives the minimum center distance for a given load in worm gearing.

Differentiating equation (A) w.r.t 'λ' and equating to zero

$$\Rightarrow \qquad \frac{(V_r)\sin^3\lambda - \cos^3\lambda}{\sin^2\lambda \cdot \cos^2\lambda} = 0 \text{ or } V_r = \cos^3\lambda \text{ (for lowest point on velocity ratio curve)}$$

3.9 FORCES ACTING ON WORM GEARS

Forces acting on the worm and worm gear pair are as follows. Normal force can be resolved into three components:

- Tangential force (W_t) on worm:

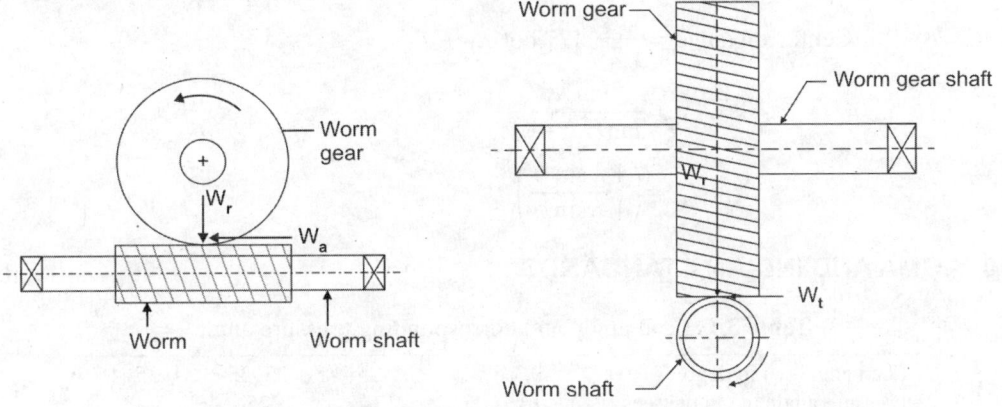

Fig. 3.9 Forces acting on worm gears

$$W_t = \frac{2 \times \text{Torque on worm}}{\text{Pitch circle diameter of worm }(d_w)} = \text{Axial force or thrust on worm gear}$$

also $\boxed{W_t = W(\cos\alpha\,\sin\gamma + \mu\,\cos\gamma)\ \text{(from geometry)}}$

[α = Normal pressure angle
γ = Lead angle
μ = Coefficient of friction]

This force generates a twisting moment ($W_t \times d_w/2$) and tends to bend worm in horizontal plane.

- Axial force or thrust (W_a) on worm

$= W_t/\tan\lambda = $ Tangential force on worm gear.

$$= \frac{2 \times \text{Torque on worm gear}}{\text{Pitch circle diameter of worm gear}\,(d_g)} = W(\cos\alpha\,\cos\gamma - \mu\,\sin\gamma)$$

[W = Normal reaction]

'W_a' tends to move worm axially and induces axial thrust on bearing and tends to bend the worm in a vertical plane,

Bending moment being '$W_a \times d_w/2$'

- Radial force (W_r) on worm:

$$W_r = W_a \tan\phi$$

W_r tends to force worm and worm gear out of mesh and bends the worm in vertical plane.

Glossary

Distance/Efficiency

• Axial pitch (P_a): It is the distance between corresponding points on adjacent teeth measured along the direction of the axis.

• Lead (L): The distance by which a helix advances along the axis of the gear for one turn around is known as lead. In a single helix, the axial pitch is equal to lead. In a double helix, this is one half of the lead; in a triple helix, this is one third of lead and so on.

• Lead angle (λ): It is the angle at which the teeth are inclined to the normal to the axis of rotation, the lead angle is the compliment of the helix angle.

$$\psi + \lambda = 90°$$

$$\text{Velocity ratio} = \frac{l}{\pi d_2} \ (d = \text{Diameter})$$

$$\text{Center distance} = \frac{m_2}{2} \ [T_1 \cot \lambda_1 + T_2]$$

$$\eta = \frac{\tan \lambda_1}{\tan (\lambda_1 + \phi)}$$

$$\eta_{max} = \left(\frac{1 - \sin \phi}{1 + \sin \phi} \right)$$

3.10 AGMA AND INDIAN STANDARDS

Table 3.1 Lead angle and corresponding pressure angle

Lead angle (λ) in degrees	0–16	16–25	25–35	35–45
Pressure angle (ϕ) in degrees	$14\frac{1}{2}$	20	25	30

Table 3.2 Preferred values of ($z_1/z_2/q/m$) for worm gears

Transmission ratio (Approx.)	Center distance (mm)				
	100	125	160	200	250
20	2/40/10/4	2/40/10/5	–	2/40/10/8	2/40/10/10
25	–	2/52/10/4	2/54/10/5	–	2/52/10/8
30	1/30/10/5	1/31/10/6	1/30/10/8	1/30/10/10	–
40	1/40/10/4	1/40/10/5	–	1/40/10/8	1/40/10/10
50	–	1/52/10/4	1/54/10/5	–	1/52/10/8

Table 3.3 Values of the surface stress factor S_c

Materials		Values of S_c when running with			
		A	B	C	D
A.	Phosphor bronze (centrifugally cast)	–	0.85	0.92	1.55
	Phosphor bronze (sand cast and chilled)	–	0.63	0.70	1.27
	Phosphor bronze (sand cast)	–	0.47	0.54	1.06
B.	0.4% carbon steel—normalized (40C8)	1.10	–	–	–
C.	0.55% carbon steel—normalized (55C8)	1.55	–	–	–
D.	Case-hardened carbon steel (10C4, 14C6)	4.93	–	–	–
	Case-hardened alloy steel (16Ni80Cr60, 20Ni2Mo25)	5.41	–	–	–
	Nickel chromium steel (13Ni3Cr80, 15Ni4Cr1)	6.19	–	–	–

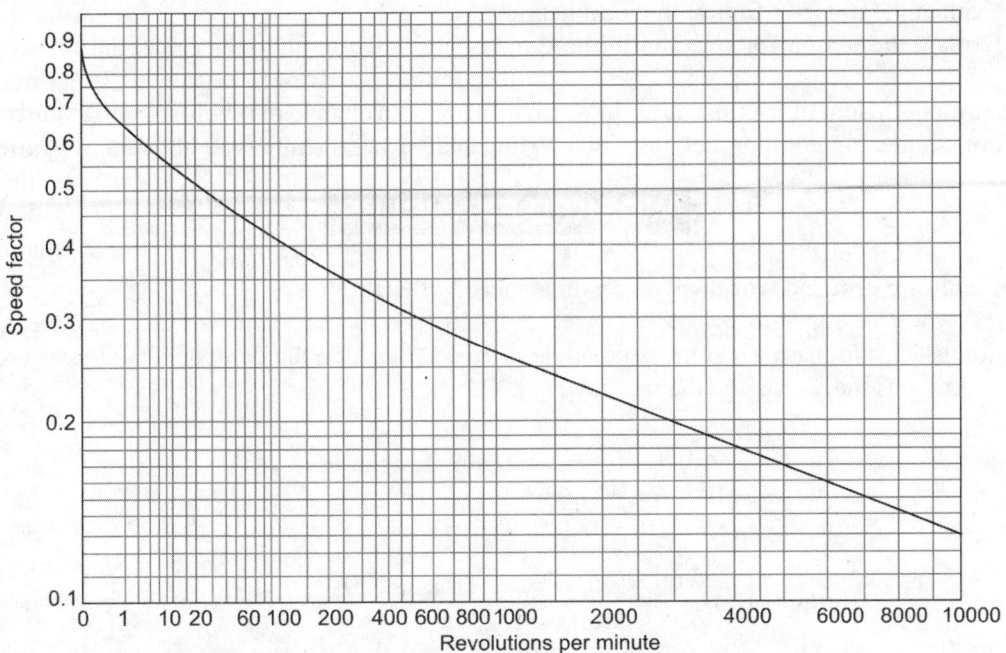

Fig. 3.10 Speed factor for worm gears for strength (X_b)

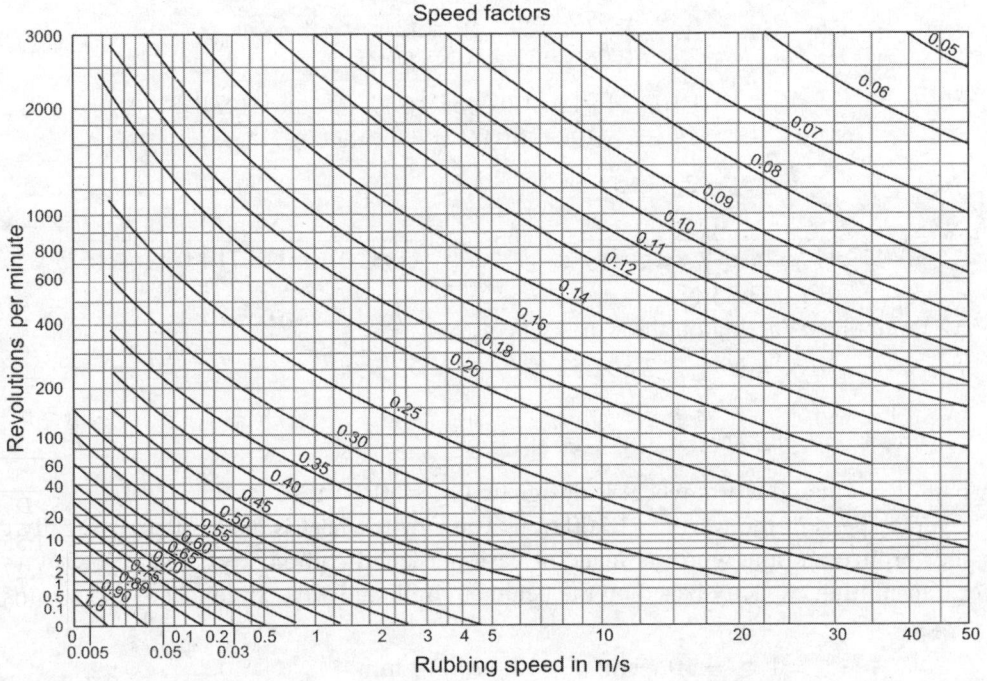

Fig. 3.11 Speed factor for worm gears for wear (X_c)

REVIEW QUESTIONS

1. Explain the mode of failure in worm gearing.
2. Explain the reason for selection of hard material for worm and softer material for worm gear.
3. How coefficient of friction varies in worm gearing with change in rubbing/sliding velocity?
4. For connecting non-parallel, non-intersecting shafts, suggest the type of gears to be used.

WORKED EXAMPLES

1. A pair of worm and worm wheel designated as 3/60/10/6.

The worm is transmitting 5 kW power at 1,440 rpm to the worm wheel. Coefficient of friction is 0.1 and the normal pressure angle is 20°. Determine the components of gear tooth force acting as the worm and the worm wheel.

SOLUTION:

$z_1 = 3$, $z_2 = 60$, $q = 10$, $m = 6$ (from designation)

also $\quad d_1 = qm = 10 \times 6 = 60$ mm

$\tan \gamma = z_1/q = 3/10 = 0.3$ or $\gamma = 16.7°$

$$\text{torque}(M_t) = \frac{60 \times 10^6 \times P_{kW}}{2\pi n_1} = \frac{60 \times 10^6 (5)}{2\pi \times 1,440}$$

$$= 33{,}157.28 \text{ N-m}$$

Now tangential force $(W)_t = \dfrac{2M_t}{d_1} = \dfrac{2 \times 33{,}157.28}{60} = 1{,}105.24\,\text{N}$

and axial force $(W)_a = W_t \times \dfrac{(\cos\alpha\cos\gamma - \mu\sin\gamma)}{(\cos\alpha\sin\gamma + \mu\cos\gamma)}$

$$= \frac{1{,}105.24 \times [\cos(20)\cos(16.7) - 0.1\sin(16.7)]}{[\cos(20)\sin(16.7) + 0.1\cos(16.7)]}$$

$$= 2{,}632.55 \text{ N}$$

$(P_1)_r = (P_1)_t \times \dfrac{\sin\alpha}{(\cos\alpha\sin\gamma + \mu\cos\gamma)} = 1{,}105.24 \times \dfrac{\sin 20}{[\cos(20)\sin(16.7) + 0.1\cos(16.7)]}$

$$= 1{,}033.35 \text{ N}$$

Force components acting on the worm wheel:

$(P_2)_t = (P_1)_a = 2{,}632.55$ N

$(P_2)_a = P_1 = 1{,}105.24$ N

$(P_2)_r = (P_1)_r = 1{,}033.35$ N.

2. A pair of worm and worm wheel is designated as 1/30/10/10.

The input speed of the worm is 1,200 rpm. The worm wheel is made of centrifugally cast, phosphor bronze and the worm is made of case-hardened carbon steel $14C_6$. Determine the power transmitting capacity based on the beam strength and wear strength.

SOLUTION:

$z_1 = 1$, $z_2 = 30$ teeth, $q = 10$, $m = 10$ mm

$i = \dfrac{z_2}{z_1} = \dfrac{30}{1} = 30$, $n_1 = 1{,}200$ rpm

So
$$n_2 = \frac{1,200}{i} = \frac{1,200}{30} = 40 \text{ rpm}$$
$$d_2 = mz_2 = 10 \times 30 = 300 \text{ mm}$$

$$\tan \gamma = \frac{z_1}{q} = \frac{1}{10} = 0.1 \Rightarrow \gamma = 5.71°$$

as
$$F = 2m\sqrt{1+q} = 2(10)\sqrt{10+1} = 66.33 \text{ mm}$$
$$C = 0.2 \ m \cos \gamma = 0.2 \ (10) \cos (5.71) = 1.99 \text{ mm}$$
$$d_{a_1} = m \ (q + 2) = 10 \ (10 + 2) = 120 \text{ m}$$

$$l_r = \left(d_{a_1} + 2C\right)\sin^{-1}\left[\frac{F}{d_{a_1} + 2C}\right]$$

$$= (120 + 2 \times 1.999)\sin^{-1}\left[\frac{66.33}{120 + 2 \times 1.99}\right]$$

$$= 69.988 \text{ mm}$$

For case-hardened carbon steel $14C_6$,
$$S_{b_1} = 28.2 \quad \text{(from table)}$$
For centrifugally cast phosphor bronze
$$S_{b_2} = 7.0$$

From graph $X_{b_1} = 0.25$ for $n_1 = 1,200$ rpm

$$X_{b_2} = 0.48 \text{ for } n_2 = 40 \text{ rpm}$$

$\Rightarrow$
$$\left(M_t\right)_1 = 17.65 X_{b_1} S_{b_1} m l_r d_2 \cos \gamma$$
$$= 17.65(0.25)(28.2)(10)(69.988)(300) \cos (5.71)$$
$$= 25,996.711 \text{ N-mm}$$
$$\left(M_t\right)_2 = 17.65 X_{b_2} S_{b_2} m l_r d_2 \cos \gamma$$
$$= 17.65(0.48)(7.0)(100)(69.998)(300) \cos (5.71)$$
$$= 12,389.922 \text{ N-mm}.$$

The lower value of the torque on the worm wheel is 12,389,922 N-m.

$\Rightarrow$
$$P_{\text{kW}} = \frac{2\pi m_2 \left(M_t\right)}{60 \times 10^6} = \frac{2\pi (40)(12,389,922)}{60} \times 10^6$$
$$= 51.9$$

Design on the basis of wear strength:

as
$$m = 10 \text{ mm}, \ d_2 = 300 \text{ mm}$$

For $(q = 10)$ and $(z_1 = 1)$, or factor Y_2 (from table) is given by
$$Y_2 = 1.143$$

For case-hardened carbon steel $14C_6$
$$S_{c_1} = 4.93$$

For centrifugally cast phosphor bronze
$$S_{c_2} = 1.55$$

So
$$V_S = \frac{\pi d_1 n_1}{60,000 \cos \gamma} = \frac{\pi (10 \times 10)(1,200)}{60,000 \cos (5.71)} = 6.315 \text{ m/sec.}$$

For $\qquad V_S = 6.315$ m/s and $n_1 = 1,200$ rpm

$\qquad X_{c_1} = 0.112$

For $\qquad V_S = 6.315$ m/s and $n_1 = 40$ rpm

$\qquad X_{c_2} = 0.26$

Now $\qquad (M_t)_3 = 18.64 X_{c_1} S_{c_1} Y_2 (d_2)^{1.8} m$

$\qquad = 18.64 \ (0.112)(4.93)(1.143)(300)^{1.8}(10)$

$\qquad = 3,383,570.4$ N-m

$\qquad (M_t)_4 = 18.64 X_{c_2} S_{c_2} Y_2 (d_2)^{1.8} m$

$\qquad = 18.64(0.26)(1.55)(1.143)(300)^{1.8}(10)$

$\qquad = 246,953.5$ N-mm

The lower value of torque on worm wheel is 2,469,535.8 N-m. Therefore,

$$P_{kW} = \frac{2\pi n_2 (M_t)}{60 \times 10^6} = \frac{2\pi(40)(2,469,535.8)}{60 \times 10^6}$$

$$= 10.34.$$

PREVIOUS YEAR QUESTION PAPERS/PROBLEMS (UPTU)

1. Design a worm gearing to transmit 11 kW from an electric motor running at 1,500 rpm to a machine running at 75 rpm. Load is intermittent (<3 hours of continuous service) and steady.

[UPTU 2007]

SOLUTION:

Given: Power transmitted (P) = 11 kW

$\qquad$ Speed of motor N_w = 1,500 rpm

$\qquad$ Speed of machine = 75 rpm

Let us assume that normal pressure angle is to be 20°. The lead angle γ shall lie between 15 to 30°.

$$\text{Velocity ratio} = \frac{1,500}{75} = 20$$

Allowing 6 degrees per thread of worm, gear could have 2 to 5 teeth.

Selecting a triple thread worm.

$\qquad T_g = 3 \times 20 = 60$

Approximate center distance

$\qquad a = 8[(i + 5) \times P]^{0.588}$

where $\qquad i$ = Gear ratio = 20

$\qquad P$ = Power transmitted in kW

$\qquad a = 8 \ [(20 + 5) \times 11]^{0.588} = 217.5$

Say $\qquad a = 218$ mm.

Assuming center distance of 200 m

now pitch diameter of worm

$\qquad d_w = (0.5 \text{ to } 0.9) \ a^{0.875}$

or $\qquad d_w = (0.5 \text{ to } 0.9)(200)^{0.875}$

$\qquad = 51.56 \text{ to } 42.8$ mm

Let us select

$\qquad d_w = 80$ m, also $d_w = 3 P_g$

From design data, selecting this module $(m) = 8$ mm

Pitch diameter of gear $(d_g) = mT_g = 8 \times 60 = 480$ mm

Actual center distance

$$a = \frac{1}{2}\left(d_w + d_g\right) = \frac{1}{2}(80 + 480) = 280 \text{ mm}$$

From table, for transmission ratio of 20 and center distance of 200 mm (as 280 mm is not standard, recommended values are 2/40/10/8)

$$\Rightarrow \qquad z_1 = 2, \ z_2 = 40, \ q = 10, \ m = 8 \text{ mm}$$
$$d_g = qz_2 = 10 \times 40 = 400 \text{ mm}$$

For two start and module 8 mm and $g = 10$, the axial pitch $(P_x) = 25.133$

Lead $\qquad P_2 = z_1 \times P_x = 2 \times 25.133 = 50.266$ mm

$$\tan \gamma = \frac{P_2}{\pi d_w} = \frac{50.266}{\pi \times 80} = 0.2$$

$$\therefore \qquad \gamma = \tan^{-1}(0.2) \text{ or } \gamma = 11.3°$$

From design table, dimensions of worm corresponding to q, r, etc. are

Tip diameter (addendum diameter) $\left(d_{a_1}\right) = 96$ mm

Root diameter (dedendum diameter) $\left(d_{f_1}\right) = 60.8$ mm

Velocity of worm $\left(V_w\right) = \dfrac{\pi d_w N_w}{60 \times 1000}$

$$V_w = \frac{\pi \times 80 \times 1,500}{60 \times 1,000}$$

$$= 6.28 \text{ m/sec}$$

Velocity of gear $\left(V_g\right) = \dfrac{\pi d_g N_g}{60 \times 1,000} = \dfrac{\pi \times 400 \times 75}{60 \times 1,000} = 1.57$ m/sec

$$F_{t_g} = \frac{1,000 \times P}{V_g} = \frac{1,000 \times 11}{1.57}, \ F_{t_g} = 7,006 \text{ N}$$

Assuming gear to be made of phosphor bronze centrifugally cast having

$$\sigma = 262.6 \text{ N/mm}^2$$

HB = 90 and work of surface hardened carbon steel $55C_6$ having $\sigma = 695.14$ N/mm^2, HB = 520, taking a factor of safety of 3,

$\sigma_0 = 87.5$ N/mm^2 for gear and 231.7 N/m^2 for worm. We shall design the gear.

$$F_{t_g} = F_d = \sigma_0 \pi m_n \text{ by } C_V$$

and $\qquad m_n = m_n \cos P_g = m_n \cos \gamma$

$$m_n = 8 \cos (4.3) = 7.85 \text{ m}$$

For $20°$ normal pressure angle, $y = 0.125$

From design data table

$$C_V = \frac{6.1}{6.1 + V_g} = \frac{6.1}{6.1 + 1.57} = 0.795$$

$$\therefore \qquad 7,006 = 87.5 \ a \times 7.85 \times b \times 0.125 \times 0.79$$

$$b = 32.68 \text{ mm}$$

So $\qquad b = 34$ m

From data table, $b = 0.75\,d_w$ or $b = 0.75 \times 80$
or face width $b = 60$ m
Let us adopt face width $(b) = 60$ mm.

Checking for Wear

$$F_w = d_g b \text{ kW}$$

For hardened steel worm (over 100 HB) and bronze gear kW = 0.55 N/mm^2
For $a_n = 20°$
$$F_w = 400 \times 60 \times 0.55 = 13,200 \text{ N}$$

as $F_w < F_{t_g}$, hence design is safe. Length of wear from Design Data Book is

$$L \geq (11 + 0.0622) \text{ m.}$$
$\Rightarrow$ $L = 107.2$ m $\Rightarrow$ 108 m.

2. A hardened steel worm rotating at 1,250 rpm transmits 12 kW to a phosphor bronze gear with a transmission ratio 15:1. The center distance is 22.5 mm and teeth have $14\frac{1}{2}°$ FD involute form. Design the drive.   (UPTU 2008)

SOLUTION:

Speed of worm   = 1,250 rpm
Power transmitted   = 12 kW
Transmission ratio   = 15:1
Velocity ratio (V_r) = 15

$\Rightarrow$ Speed of gear $= \dfrac{1,250}{15} = 83.33$ rpm

Center distance $(x) = 225$ mm
Pressure angle $= 14\frac{1}{2}$ °C full depth involute

Worm Design

If l_N = Normal lead and λ = Lead angle.
   From graph, as value of x/l_N will be minimum corresponding to $\cot^3\lambda = V_r = 15$
$$\Rightarrow \lambda = 22.1°$$

also $\dfrac{x}{l_N} = \dfrac{1}{2\pi}\left(\dfrac{1}{\sin\lambda} + \dfrac{V_r}{\cos\lambda}\right)$

$\Rightarrow$ $\dfrac{225}{l_N} = \dfrac{1}{2\pi}\left[\dfrac{1}{\sin(22.1)} + \dfrac{5}{\cos(22.1)}\right] = \dfrac{1}{2\pi}(2.66 + 16.2) = 3$

So $l_N = \dfrac{225}{3} = 75$ mm

and axial lead $l = \dfrac{l_N}{\cos\lambda} = \dfrac{75}{\cos(22.1)} = 80.95\,\text{mm}$

   For velocity ratio of 15, number of starts or threads on the worm $(n) = T_w = 2$
   Axial pitch of threads on worm $(P_a) = l/2 = 80.95/2$

$$P_a = 40.47 \text{ mm}, \quad m = \dfrac{P_a}{\pi} = 12 \text{ mm}$$

Axial pitch of threads on worm $(P_a) = \pi m = \pi \times 12 = 37.68$ mm
and axial lead of thread on worm

$$l = P_a \times n = 37.62 \times 2 = 75.36 \text{ mm}$$

Normal lead of thread on worm

$$l_N = l \cos \lambda = 80.95 \cos (22.1)$$
$$l_N = 75 \text{ mm}$$

also center distance $(x) = \dfrac{l_N}{2\pi}\left(\dfrac{1}{\sin \gamma} + \dfrac{V_r}{\cos \alpha}\right)$

$$x = \frac{75}{2\pi}\left(\frac{1}{\sin(22.1)} + \frac{15}{\cos(22.1)}\right) = 225 \text{ mm}$$

$$x = 225 \text{ mm} \Rightarrow \text{(result is correct)}$$

Let $\qquad d_w$ = Pitch circle diameter of worm
also $\qquad \tan \lambda = l/\pi d_w$

$$d_w = \frac{l}{\pi \tan \gamma} = \frac{80.95}{3.14 \times \tan(22.1)} = 63.48$$

$$d_w = 64 \text{ mm.}$$

Since velocity ratio is 15 and worm has double threads ($n = T_w = 2$), therefore, number of teeth on worm gear $(T_g) = 15 \times 2 = 30$

We find that face length of worm or length of threaded portion $(L_w) = P_c (4.5 + 0.2 \times T_w) = 37.68 (4.5 + 0.2 \times 2) = 171$ mm $(P_c = P_a)$

This length should be increased by 25 to 30 mm for the feed marks produced by vibrating grinding wheel as it leaves thread root,

So taking $L_w = 195$ mm
also depth of tooth $(h) = 0.623\, P_c = 0.623 \times 37.68 = 23.47$ mm.

addendum $(a) = 0.286 P_c = 10.72$ mm
Outside diameter of worm $(d_{ow}) = d_w + 2a = 63.48 + 2 \times 10.72$
So $\qquad d_{ow} = 84.91$ mm

Design of Worm Gear

Pitch circle diameter of worm gear

$$d_g = mT_g = 12 \times 30 = 360 \text{ mm.}$$

Outside diameter of worm gear

$$d_{og} = d_g + 0.8903\, P_c = 360 + 0.8903 \times 37.68$$
$$d_{og} = 393.6 \text{ mm.}$$

Throat diameter

$$d_T = d_g + 0.572\, P_c = 360 + 0.572 \times 37.68$$
$$d_T = 381.6 \text{ mm.}$$

and face width $\qquad b = 2.15\, P_c + 5$ mm
or $\qquad b = 2.15 \times 37.68 + 5 = 86$ mm

Checking of Designed Worm Gearing as per Tangential Load, Dynamic Load

a. Checking for tangential load, let N_g = Speed of worm gear in rpm.

as velocity ratio $(V_r) = \dfrac{N_w}{N_g}$ or $N_g = \dfrac{N_w}{V_r} = \dfrac{1,250}{15} = 83.33$ rpm

torque transmitted $(T) = \dfrac{P \times 60}{2\pi N_g} = \dfrac{12,000 \times 60}{2\pi \times 83.333}$

$$T = 1,376 \text{ N-m}$$

$$\text{Tangential load} = \dfrac{2 \times \text{Torque}}{d_g} = \dfrac{2 \times 1,375}{0.394} = 6,983 \text{ N}$$

Also pitch line or peripheral velocity of worm gear

$$V = \dfrac{\pi d_g N_g}{60} = \dfrac{\pi \times 0.394 \times 83.33}{60} = 1.7182 \text{ m/s}$$

Velocity factor $(C_V) = \dfrac{6}{6+V} = \dfrac{6}{6+1.7182} = 0.777$

Tooth form factor for $14\frac{1}{2}°$ involute teeth

$$y_g = 0.124 - \dfrac{0.684}{T_g} = 0.124 - \dfrac{0.684}{30}$$

$$y_g = 0.10102$$

Material for worm gear is phosphor bronze whose allowable static stress is:
$$\sigma_0 = 843 \text{ MPa or } \sigma_0 = 84 \text{ N/mm}^2$$

Design tangential load:
$$W_t = (\sigma_0 C_v)b \; \pi m y = 84 \times 0.77 \times 86 \times 3.14 \times 12 \times 0.10102 = 21,211 \text{ N.}$$

Since this is more than the tangential load acting on the gear (i.e. 6,983 N), therefore, this design is safe from the point of view of tangential load.

b. Checking for dynamic load:

$$W_d = \dfrac{W_t}{V} = \dfrac{21,211}{1.7182} = 12,344.8 \text{ N}$$

Since this is more than $W_t = 6,983$ N, therefore, the design is safe from dynamic load resistance point of view.

c. Checking for static load or endurance strength:

We know that the flexural endurance limit for phosphor bronze is
$$\sigma = 168 \text{ MPa}$$

Static load $(W_s) = \sigma = b\pi m y_g$

$$W_s = 168 \times 86 \times 3.14 \times 12 \times 0.10102$$
$$= 55,093 \text{ N}$$

Since this is much more than $W_t = 6,983$ N, therefore, the design is safe from static load point of view.

d. Checking for wear:

Material for worm is given, it is hardened steel, therefore, for hardened steel worm and phosphor bronze worm gear, value of load stress factor is
$$K = 0.55 \text{ N/m}^2$$

$\therefore$ Limiting a maximum load for wear
$$W_w = d_g b \times K = 360 \times 86 \times 0.55$$
$$W_w = 17,028 \text{ N}$$

Since this is more than $W_t = 6,983$ N, therefore, design is safe from wear point of view.

4

Sliding Contact Bearings

REVISION CHECKLIST

Essential Points

- Types and selection criteria
- Plain journal bearing
- Hydrodynamic lubrication
- Properties and materials
- Lubricants and lubrication
- Hydrodynamic journal bearing
- Heat generation
- Design of journal bearing
- Thrust bearing—pivot and collar bearing
- Hydrodynamic thrust bearing

4.1 INTRODUCTION

Sliding contact bearing refers to relative motion between the rubbing surfaces of bodies such as shaft and the housing, by means of lubrication with minimum friction.

Bearing ensures free rotation of shaft; support of shaft; holding shaft in correct position; absorbing the force acting on shaft or axle and transmitting to frame on foundation. Depending upon types of friction between shaft and the bearing surface, bearings are either 'sliding' or 'contact'. Sliding contact bearing or journal bearings involve surface of the shaft sliding over the surface of the bush resulting in friction and wear. To reduce friction between the two surfaces, these are separated by a film of lubricating oil.

These find application in some of the following applications:

1. Centrifugal pumps   2. Large electric motors
3. Concrete mixers   4. Steam and gas turbines
5. Rope conveyors

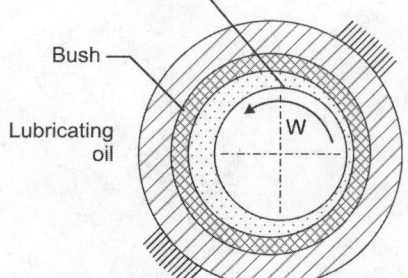

Fig. 4.1 Sliding contact bearing

4.2 TYPES OF BEARING

Sliding Contact

Sliding contact bearings are classified on the basis of lubrication between rubbing surfaces of bodies having relative motion, i.e. shaft and bearing.

On this basis sliding contact bearings are classified as **Thick Film:** 'Hydrodynamic' and 'Hydrostatic' lubrication bearings, **Thin Film** and **Zero Film** lubrication.

Thick film lubrication refers to a thick film of lubricant completely separating two surfaces of bearing in relative motion. On this criteria sliding contact bearings are: Hydrodynamic and Hydrostatic lubrication bearings.

Hydrodynamic lubrication refers to the system of lubrication in which load supporting fluid film is created by the shape and relative motion of the sliding surfaces.

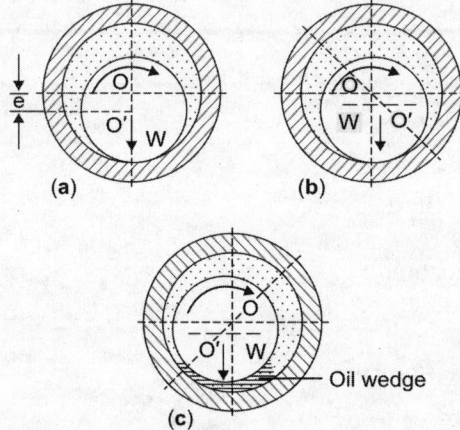

Fig. 4.2 Hydrodynamic lubrication (sliding contact). (a) Journal at rest
(b) Journal starts to rotate (c) Journal at full speed

This pressure distribution (as shown above) is created due to rotation of the shaft within the system, and hence this type of bearing is also known as self-acting bearing.

Hydrodynamic lubrication finds application in bearings mounted on "centrifugal pumps and engines", and the corresponding bearing is termed as "journal bearing" which supports the load in radial direction and portion of shaft inside bearing being termed as "journal".

Hydrodynamic journal bearings (sliding contact) are further classified as full journal and partial bearings.

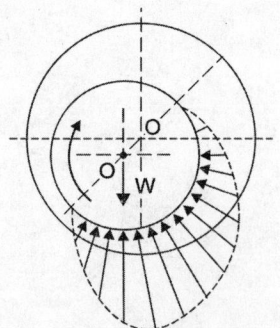

Fig. 4.3 Pressure distribution
(Hydrodynamic lubrication)

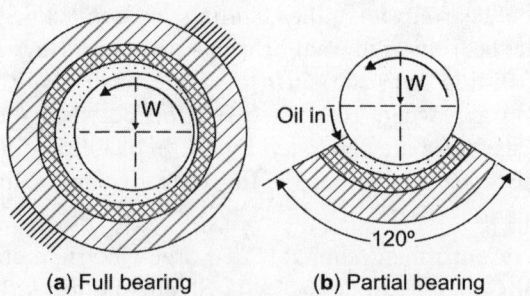

(a) Full bearing   **(b) Partial bearing**

Fig. 4.4 Full and partial bearings

Full journal bearing have a 360° angle of contact between bearing and the journal and can take load in any radial direction. These find application in most of the bearings used in industrial work.

Partial bearing comprises an angle of contact between journal and bearing less than 180° and in practice, normally around 120°. These can take load only in one radial direction and find application in rail road cars, and have simple construction, less functional loss and easy lubrication.

A full bearing is also termed as 'clearance bearing' in which radius of 'journal' is less than radius of bearing, whereas a partial bearing is termed as 'fitted bearing' in which the radius of journal and bearing are equal.

For axial loads, there are two types of thrust bearing: Footstep and collar.

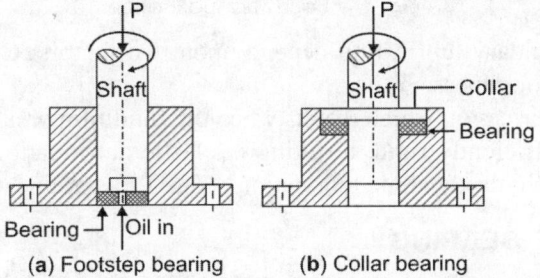

(a) Footstep bearing   (b) Collar bearing

Fig. 4.5 (a) Footstep and (b) Collar bearing

Footstep bearing is a 'thrust bearing' having a surface contact between end of the shaft and bearing surface whereas 'collar bearing' is a thrust bearing in which a collar integral with the shaft is in contact with the bearing surface.

Sliding contact bearings are also classified under hydrostatic lubrication which comprises load supporting fluid film separating shaft (journal) and bearing is created by an external source (say) pump, supplying sufficient fluid under pressure and is thereby also termed as externally pressurized bearing. These find application in centrifuges, vertical turbo generators and ball mills.

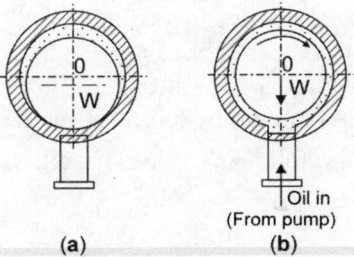

(a)   (b)

Fig. 4.6 Hydrostatic lubrication (sliding contact). (a) Journal at rest
(b) Journal at full speed

Hydrostatic lubrication sliding contact bearings are simple in construction and have high load carrying capacity even at low speeds (in contrast to hydrodynamic ones). These have no starting friction and slipping/rubbing action at any speeds. Maintenance costs are also low for these type of bearings.

Third classification is thin film lubrication (sliding contact) which has a relatively thin lubricant film and a partial metal to metal contact between journal (shaft) and the bearing. These find application in door hinges and machine tool slides.

This basically results from excessive load, low speed, insufficient surface area or oil supply and misalignment. This is also termed as 'boundary lubrication'.

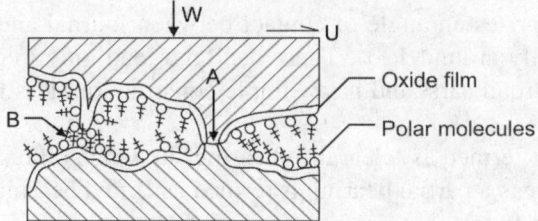

A—Metal to metal contact and B—Cluster of molecules

Fig. 4.7 Boundary lubrication

Bearings under boundary lubrication depend upon two factors: Chemical composition of lubricant and surface roughness.

Another mode of lubrication which results when the fluid film pressure is high and surfaces to be separated not sufficiently rigid, resulting in elastic deformation of contacting surfaces and is termed as 'elastohydrodynamic lubrication'. These find application in gears, wears, etc.

4.3 SELECTION OF BEARING

Bearing selection is done on the basis of following criteria (for sliding contact).

- Load bearing—good
- Capacity—power consumption varies as $n^2 d^3 l$ (n—rpm, d—diameter, l—length)
- Friction (starting)—poor
- Space requirements—small
- Accuracy—good
- Noise—quiet
- Cost—small in mass production
- Type of failure—permits limited emergency operation after failure
- Damping of vibrations—good
- Life—unlimited (except in case of cyclic loading)
- Maintenance—requirement of clean lubricant
- Low temperature—poor operation
- High temperature operation—limited by lubricant (used)

On the basis of above parameters, following values need to be determined.

- l/d ratio (length to diameter ratio)
- Unit bearing pressure
- Start up load
- Radial clearance
- Minimum oil film thickness
- Maximum oil film temperature

4.4 PLAIN JOURNAL BEARING

Refer Sections 4.1 and 4.2 (Journal Bearing)

4.5 HYDRODYNAMIC LUBRICATION

Refer Section 4.2 (Hydrodynamic Lubrication)

4.6 PROPERTIES AND MATERIALS

Bearing (sliding contact) materials are selected on the basis of operating factors like:

- Load
- Speed
- Life
- Temperature
- Type and quantity of lubricant available
- Contamination

Following properties of materials on the basis of above operating factors are taken into account.

- Compatibility: A good bearing material should not weld easily to the journal material. Welding may occur due to metal to metal contact between journal and bearing.
- Conformability: Bearing material should be able to compensate for misalignment of journal and other errors due to creep or plastic deformation.
- Material should be soft with a low modulus of elasticity.
- Embedability: Bearing material should be able to absorb external contaminants like dust in lubricant to avoid wear and scoring.
- Fatigue resilience: Fatigue resilience should be sufficient so as to withstand repeated loads without developing surface fatigue cracks. Important in aircraft and automotive engines.
- Corrosion resistance: Should be good. Important in applications like IC engines, can be done by applying a thin layer of indium over cadmium or lead alloys.
- Thermal conductivity: Should be good so as to allow rapid removal of heat produced due to friction.
- Thermal expansion: Coefficient of thermal expansion and effect on clearance should be considered, if bearing is to operate at high temperature.
- Load capacity: Material should be of high compressive strength to withstand maximum hydrodynamic pressure so as to avoid its extrusion.
- Bondability: For high capacity applications, bearings are made by bonding one or more thin layers of bearing material to a high strength steel shell. So material should be bondable.

Bearing Materials

Due to good fatigue strength, high load capacity, good high temperature performance; copper lead materials are extensively used for heavy duty main and connecting rod bearings for automobiles, trucks and aircrafts.

For intermediate loads and speeds, lead bronzes are used as bearing material in machine tools, home appliances, farm machinery and pumps.

For high load, low speed applications, tin and bronzes are used, e.g. in earth moving machinery, i.e. engines for connecting rod bearings, etc.

For relatively light loads, cast iron and steel bearings may also be used.

4.7 LUBRICANTS AND LUBRICATION

Purpose of lubricant is to:

a. Reduce friction losses and wear between rubbing surfaces.

b. Carry off heat of friction.

c. Provide protection against corrosion.

On these criteria, lubricants are classified into following three types.

• Fluids/oils
• Greases
• Solid film lubricants

a. Oils: These are slippery hydrocarbon liquids. Synthetic oils can operate over a wide range of temperature with little change in viscosity.

For example, polyalkylene glycols and silicones,

These are easy to drain and refill and are indispensable for dissipation of large amount of heat from the bearing housing; also these lubricants readily feed into all areas of contact and can carry away dirt, water and wear products.

This class of lubricants are available for a greater range of operating speeds and temperature than grease.

b. Grease: Grease is a semisolid combining a fluid lubricant with a thickening agent using a soap comprising calcium, sodium and lithium.

These can be easily retained in a housing and contained. So leakproof designs are unnecessary. These are employed where slow speed and heavy pressure exist.

• Less maintenance is required as there is no oil lever to be maintained.
• Has better sealing abilities than oils keeping dirt and moisture out of housing.

c. Solid film lubricants: These are useful in reducing friction where oil films cannot be maintained because of high pressure or temperatures. They should be softer than the materials being lubricated. Graphite is a most common solid lubricant used either alone or mixed with oil and grease.

Lubricants should be chemically stable with bearing material and atmosphere at all temperatures encountered in the application along with a little change in viscosity with change in temperature.

Society of Automotive Engineers (SAE) has classified lubricating oils by a number related to viscosity of oil and are classified as follows:

Table 4.1 Bearing lubricants

| SAE number | Saybolt universal viscosity at 210°C | |
Minimum	Minimum	Maximum
20	45	Less than 58
30	58	Less than 70
40	70	Less than 85
50	85	Less than 110

Here (as can be seen from table), SAE number corresponds to approximately one half of the viscosity of oil at 210 °F. When SAE number is more, it indicates more viscous oil.

4.8 HYDRODYNAMIC JOURNAL BEARING

Refer Section 4.2 (Hydrodynamic Journal Bearing)

4.9 HEAT GENERATION

Heat generation in sliding contact bearing occurs due to viscosity of the lubricating oil. Frictional work done is converted to heat, which increases the temperature of the lubricant.

If we assume that the total heat generated in the bearing is carried away by the total (lubricant oil) flow in the bearing, temperature rise can be determined.

So power (in kilowatts)

$$P_{kW} = (2\pi n_s)\left(f_{W_r}\right)(10^{-6})$$

(n_s = Journal speed (rev/sec); f is the friction factor or coefficient of friction, W is axial load, r is radius of journal)

So the heat generation H_g is:

$$H_g = P_{kW} = (2\pi n_s)\left(f_{W_r}\right) \times 10^{-6} \text{ kW}$$

also for journal bearing, coefficient of friction 'f' is given by:

$$f = \left(\frac{C}{r}\right)(CFV) \text{ and } W = 2Plr$$

where
P = Tangential frictional force

l = Length of bearing

C = Radial clearance (between journal and bearing)

CFV = Coefficient of friction variable

$$(CFV) = \left(\frac{r}{C}\right)f$$

So
$$\boxed{H_g = (4\pi)\left(10^{-6} r C n_s l_P (CFV)\right)}$$

Heat carried away by oil from H_c is given by:

$H_c = mc_p\Delta t$

where
m = Mass of lubricant passing through bearing (in kg/sec)

C_p = Specific heat of lubricant (kJ/kg °C)

Δt = Temperature rise

mass 'm' of the lubricating oil is given by:

$m = \rho Q \times 10^{-6}$ kg/sec

where Q = Flow of the lubricant (on discharge) [mm^3/sec]

$Q = rCn_s l$ (FV)

where
FV = Flow variable $= \dfrac{Q}{rCn_s l}$

So
$m = \rho \, (rCn_s l)(FV)(10^{-6})$ kg/sec.

So
$H_c = C_p \Delta t \rho (rCn_s l)(FV)(10^{-6})$

equating:
$H_g = H_c$

$\Rightarrow$
$$\boxed{\Delta t = \frac{(4\pi P)}{\rho C_p} \frac{(CFV)}{(FV)}}$$

Average temperature of lubricant, when 'T_1' is the inlet temperature, is:

$$T_{av} = T_1 + \left(\frac{\Delta t}{2}\right)$$

(For most lubricants: $\rho \approx 0.86$ and $C_p = 1.76$ kJ/kg °C)

WORKED EXAMPLES

1. Why is the journal bearing design carried out using non-dimensional numbers?

SOLUTION: Coefficient of friction in the design of bearings is of great importance, because it afford a means for determining the loss of power due to bearing friction. Various analyses have shown that the journal coefficient of friction is a function of at least three dimensionless parameters.

(1) *ZN/p*	(2) *d/C*	(3) *l/d*
(Bearing	(Diameter to	(Length to
characteristic no.)	clearance ratio)	diameter ratio)

Here '*Z*' is the absolute viscosity at its operating temperature (kg/ms).

'*N*' is the speed of journal (rev/min).

'*p*' is the bearing pressure based on projected area (N/m^2).

'*C*' is the diametral clearance between journal and bearings (in m).

'*l*' is the length of bearing.

With the help of above mentioned non-dimensional parameters:

- Designs and simulation can be done by changing different variables for minimum coefficient of friction and reduce power losses. For example, bearing characteristic no. helps to predict performance of bearing (sliding contact) in terms of bearing lubrication.

2. With the help of bearing modulus–coefficient of friction curve, explain the stability of lubricating oil film in journal bearing.

SOLUTION:

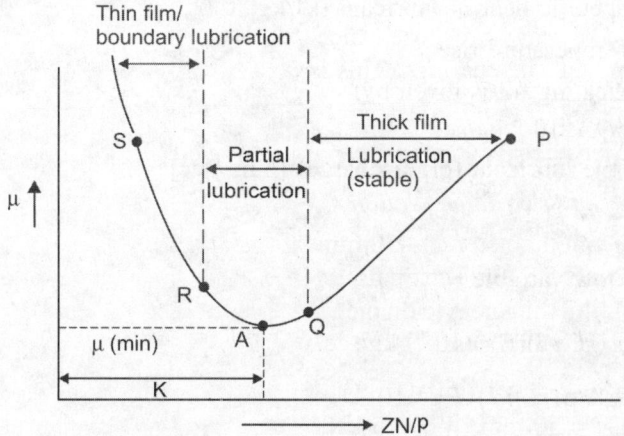

Variation of 'μ' with 'ZN/p'

From the curve (above), it can be seen that minimum friction occurs at A and at this point the value of 'ZN/p' (bearing characteristic no./modulus) is denoted by 'K'.

Bearing should not be operated at this value of bearing modulus, because of slight decrease in speed or slight increase in pressure will break the oil film and make the journal to operate with metal to metal contact. This will result in high friction, wear and heating. In order to prevent such conditions, bearing should be designed for a value of 'ZN/p', at least three times of minimum value of bearing modulus 'K'.

If bearing is subjected to large fluctuations of load and heavy impacts, value of $ZN/p = 15K$ may be used.

From the Graph, it concludes that for $\dfrac{(ZN)}{p} \gg K_1$ bearing will operate with thick film lubrication whereas for $\dfrac{(ZN)}{p} < K$, oil (lubricant) film rupture and there is a metal to metal contact.

3. Define the term: Critical bearing modulus; minimum film thickness; Summerfield number; operating oil temperature; oil bath bearings as applied to journal bearings.

SOLUTION: Refer Sections 4.10.1, 4.10.2 and 4.10.3.

Oil bath bearing is a lubrication system in which a part of journal is actually immersed in the lubricant would provide good circulation.

4. Clearly differentiate between thin film lubrication and thick film lubrication. Give examples.

SOLUTION: In thin film lubrication (imperfect), there exists an unstable condition and the metal surface may, therefore, touch each other from time to time under certain condition such as low rubbing/sliding speed or high unit loads. Thin film lubrication may continue indefinitely.

When there is no metal to metal contact, lubricant is termed as thick film lubrication, i.e. thick film bearings are those in which the (perfect) working surfaces are completely separated from each other by the lubricant.

5. Distinguish between "hydrodynamic bearings" and "hydrostatic bearings".

SOLUTION: 'Hydrodynamic bearings' are also called thick film bearings. Thick film bearings are those in which the working surfaces are completely separated from each other by the lubricant.

Whereas 'hydrostatic bearings' are those which can support steady loads without any relative motion between journal and the bearings. This is achieved by forcing externally pressurized lubricant between members.

PREVIOUS YEAR UNIVERSITY QUESTIONS

[Refer Section 4.10, Design Data Handbook by Sadhu Singh]

1. A sleeve bearing 50 mm diameter and 50 mm long has a journal speed of 3,000 rpm. The radial load on the bearing is 5.5 kN. The oil used is SAE 10 at an average temperature of 60 °C. If the ratio of minimum film thickness to diametral clearance is 0.5, determine radial clearance, heat loss and minimum film thickness. (UPTU 2010)

SOLUTION:

Given: $d = 50$ mm, $l = 50$ mm, $W = 5.5$ kN, $N = 3,000$ rpm

          $z = 19$ Pas (for 10 SAE at 60 °C)   [from 'ASME and Indian standards' section]

Also refer 'Design Data Handbook' by Sadhu Singh.

So unit bearing pressure $(p) = \dfrac{W}{ld} = \dfrac{5.5 \times 10^3}{50 \times 50} = 2.2\,\text{N/mm}^2$

$$\frac{l}{d} = \frac{50}{50} = 1$$

From the table (for $l/d = 1$): $1 - \dfrac{h_0}{C_1} = 0.6 \Rightarrow 1 - \dfrac{2h_0}{C} = 0.6$ or $\dfrac{h_0}{C} = 0.5$

(where C_1 = Radial clearance and C = Diametral clearance)

So $C = 0.15$

and $C_1 = C/2$

So coefficient of friction $(\mu) = \dfrac{33}{10^8}\left(\dfrac{ZN}{p}\right)\left(\dfrac{d}{C}\right) + K$

$$= \left(\dfrac{33}{10^8}\right)\left(\dfrac{14 \times 10^{-3} \times 3{,}000}{2.2}\right)\left(\dfrac{25}{15}\right)$$

$$= 105 \times 10^{-7}$$

L_0 = Minimum oil film thickness = 0.15×15 mm.

Rubbing/sliding speed $(U) = \dfrac{\pi d N}{60 \times 10^3} = \dfrac{\pi \times 50 \times 3{,}000}{60 \times 1{,}000} = 7.85$

So $\qquad H_g = 105 \times 10^{-7} \times 5.5 \times 7.85 = 4533 \times 10^{-7}$ (Heat generated)

2. A full journal bearing of diameter 80 mm and length 120 mm is to support a load of 20 kN at the shaft speed of 1,500 rpm. The bearing temperature is to be limited to 75 °C and the ambient room temperature is 38 °C. The viscosity of oil used is 0.0088 kg/m-s at 115 °C. Check if artificial cooling is required and find the amount of artificial heating. [UPTU 2011]

SOLUTION: Given: d = 80 mm, $\qquad l$ = 120 mm

Bearing pressure $\qquad\qquad (P) = \dfrac{W}{A} = \dfrac{W}{ld} = \dfrac{20{,}000}{120 \times 80} = 2.08\,\text{N/mm}^2$

Coefficient of friction $\qquad (\mu) = \dfrac{33}{10^8}\left(\dfrac{ZN}{p}\right)\left(\dfrac{d}{C}\right) + 0.002$

$$= \dfrac{33}{10^8}\left(\dfrac{0.0088N}{2.08}\right)\left(\dfrac{80}{0.1}\right) + 0.002$$

$$= \dfrac{111.69}{10^8}N + 0.002$$

Heat generated $\qquad\qquad (H_g) = \mu W V$

$$= \mu W \left(\dfrac{\pi d N}{60}\right)\text{J/sec}$$

$$= \left[\dfrac{111.69}{10^6}N + 0.002\right] \times 20{,}000 \times \left(\dfrac{\pi \times 0.08N}{60}\right)$$

$$= \dfrac{9{,}356.919}{10^8}N^2 + 0.168\,N$$

$$= \dfrac{9{,}356.919 \times 1{,}500}{10^8} + 0.168 \times 1{,}500 = 252.14\text{ J/sec}$$

$t_b = 75\,°C$ and $t_a = 38\,°C$, t_b = Bearing surface temperature

$$(t_b - t_a) = (t_b - t_a) = \dfrac{1}{2}(t_0 - t_a) = \dfrac{1}{2}(75 - 38) = 18.5\,°C$$

From Design Data Handbook (By Sadhu Singh)

at $\qquad$ 18.5 °C, $C = 280$ W/m²/°C

So heat dissipated $\qquad (Q_d) = CA(t_b - t_a)$
$$= 280 \times 0.120 \times 0.08 \times (75 - 38)$$
$$= 99.46 \text{ J/sec.}$$

So amount of artificial heat required
$$= 252.14 - 99.46$$
$$= 152.68 \text{ J/sec.}$$

3. Briefly explain the following:
- i. Wedge film lubrication
- ii. Squeeze film lubrication
- iii. Hydrostatic lubrication

(UPTU 2010)

SOLUTION:

i. Wedge film lubrication

The load carrying ability of a wedge film journal bearing results when the journal or the bearing rotates relatively to the load. The most common case is that of a steady load, a fixed (non-rotating) bearing and a rotating journal.

Figure (a) below shows a journal at rest with metal to metal contact at A on the line of action of the supported load. When the journal rotates slowly in the anticlockwise direction as shown in Fig. (b), the point of contact will move to B, so that the angle of AOB is the angle of sliding friction of the surface in contact at B. In the absence of a lubricant there will be dry metal to metal friction. If a lubricant is prudent in the clearance space of the bearing and journal, then a thin absorbed film of the lubricant may partly separate the surface, but a continuous fluid film completely separating the surface will not exist because of slow speed.

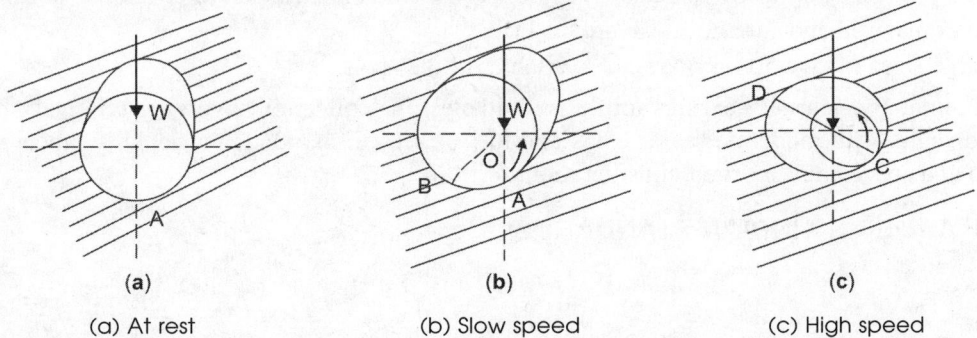

(a)	(b)	(c)
(a) At rest	(b) Slow speed	(c) High speed

When the speed of the journal is increased, a continuous fluid film is established as in Fig. (c). The center of the journal has moved so that minimum film thickness is at (c). It may be noted that pan D to C in the direction of motion, the film is continuously narrowing and hence is a converging film. The curved converging film may be considered as a wedge shaped film of a shipper being wrapped around the journal. A little consideration will show that pan C to D in the direction of rotation as shown in Fig. (c), the film is diverging and cannot give rise to a positive pressure or a supporting action.

ii. Squeeze film lubrication

In wedge film lubrication, the bearing carries a steady load and the journal rotates relatively to the bearing. But in certain cases, the bearings oscillate or rotate so slowly that the wedge film cannot provide a satisfactory film thickness. If the load is uniform or varying in magnitude while acting in a constant direction may this becomes a thin film or possibly to zero film problem. But, if the load reverses its direction, the squeeze film may develop sufficient capacity to carry the dynamic loads without contact between the journal and the bearing. Such a lubrication is called squeeze film lubrication.

iii. Hydrostatic lubrication

In hydrostatic bearings, the load supporting high pressure fluid film is created by an external source like pump. The lubricant which is pressurized externally is supplied between two surfaces. So unlike the hydrodynamic bearings, hydrostatic bearings don't require motion of one surface relative to another.

4. What are the advantages and disadvantages of journal bearings? (UPTU 2011)

SOLUTION:

Some of the advantages are:
- Low cost
- Quiet operation with little noise
- High speed capability
- They can be designed to operate with lubricants other than oil or grease for example, water or even air (dry)

Some of the disadvantages are:
- Relatively low radial load carrying capacity
- Zero thrust load capability (unless a flange bearing is used and the shaft designed with a step)
- Low misalignment capability (self-aligning types are available in small sizes, but these require the misalignment to be taken up between the outer bearing surface and the housing.)
- Shaft material and surface finish are critical.
- Large sizes (above 50 mm) are not available off the shelf.

In summary, journal bearings are most suitable for application involving relatively high speed shafts with moderate radial loads and low or zero thrust loads particularly when cost, noise and space are important considerations.

4.10 ASME AND INDIAN STANDARDS

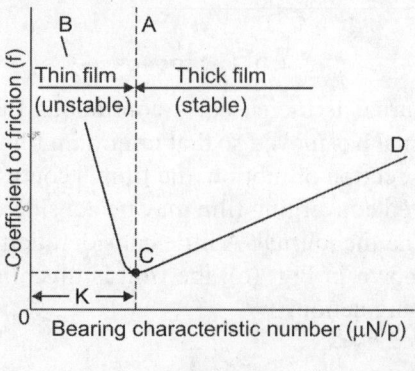

μN/p curve

Fig. 4.8 Bearing modulus (ZN/p) vs friction coefficient (f)

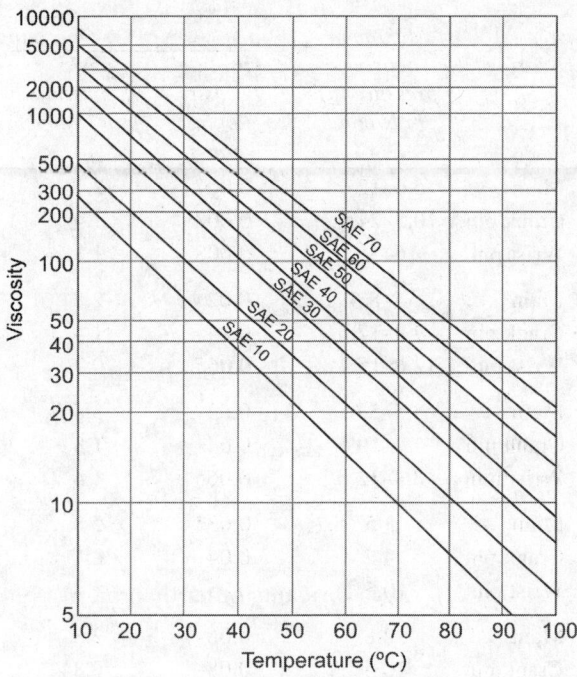

Fig. 4.9 Viscosity–temperature relationship

Table 4.2 Absolute viscosity in kg/m-s at temperature in °C

Sl. No.	Type of oil	30	35	40	45	50	55	60	65	70	75	80	90
1.	SAE10	0.05	0.036	0.027	0.0245	0.021	0.017	0.014	0.012	0.011	0.009	0.008	0.0055
2.	SAE20	0.069	0.055	0.042	0.034	0.027	0.023	0.020	0.017	0.014	0.011	0.010	0.0075
3.	SAE30	0.13	0.10	0.078	0.057	0.048	0.040	0.034	0.027	0.022	0.019	0.016	0.010
4.	SAE40	0.21	0.17	0.12	0.096	0.78	0.06	0.046	0.04	0.034	0.027	0.22	0.013
5.	SAE50	0.30	0.25	0.20	0.17	0.12	0.09	0.076	0.06	0.05	0.038	0.034	0.020
6.	SAE60	0.45	0.32	0.27	0.20	0.16	0.12	0.09	0.072	0.057	0.046	0.040	0.025
7.	SAE70	0.10	0.09	0.45	0.31	0.21	0.165	0.12	0.087	0.067	0.052	0.043	0.033

Table 4.3 Design data for journal bearings

Machinery	Bearing	Maximum bearings pressure (p) in N/mm^2	Operating values			
			Absolute viscosity (Z) in kg/m-s	ZN/p Z in kg/m-s p in N/mm^2	$\dfrac{C}{d}$	$\dfrac{l}{d}$
Automobile and aircraft engines	Main	5.6–12	0.007	2.1	–	0.8–1.8
	Crank pin	10.5–24.5	0.008	1.4	–	0.7–1.4
	Wrist pin	16–35	0.008	1.12	–	1.5–2.2
Four stroke—gas and oil engines	Main	5–8.5	0.02	2.8	0.001	0.6–2
	Crank pin	9.8–12.6	0.04	1.4	–	0.6–1.5
	Wrist pin	12.6–15.4	0.065	0.7	–	1.5–2
Two stroke—gas and oil engines	Main	3.5–5.6	0.02	3.5	0.001	0.6–2
	Crank pin	7–10.5	0.04	1.8	–	0.6–1.5
	Wrist pin	8.4–12.6	0.065	1.4	–	1.5–2
Marine steam engines	Main	3.5	0.03	2.8	0.001	0.7–1.5
	Crank pin	4.2	0.04	2.1	–	0.7–1.2
	Wrist pin	10.5	0.05	1.4	–	1.2–1.7
Stationary, slow speed steam engines	Main	2.8	0.06	2.8	0.001	1–2
	Crank pin	10.5	0.08	0.84	–	0.9–1.3
	Wrist pin	12.6	0.06	0.7	–	1.2–1.5
Stationary, high speed steam engines	Main	1.75	0.015	3.5	0.001	1.5–3
	Crank pin	4.2	0.030	0.84	–	0.9–1.3
	Wrist pin	12.6	0.025	0.7	–	1.3–1.7
Reciprocating pumps and compressors	Main	1.75	0.03	4.2	0.001	1–22
	Crank pin	4.2	0.05	2.8	–	0.9–1.7
	Wrist pin	7.0	0.08	1.4	–	1.5–2.0
Steam locomotives	Driving axle	3.85	0.10	4.2	0.001	1.6–1.8
	Crank pin	14	0.04	0.7	–	0.7–1.1
	Wrist pin	28	0.03	0.7	–	0.8–1.3
Railway, cars	Axle	3.5	0.1	7	0.001	1.8–2
Steam turbines	Main	0.7–2	0.002–0.016	14	0.001	1–2
Generators, motors, centrifugal pumps	Rotor	0.7–1.4	0.025	28	0.013	1–2
Transmission shafts	Light	0.175	0.025–0.060	7	0.001	2–3
	Fixed self-aligning	1.05	–	2.1	–	2.5–4
	Heavy	1.05	–	2.1	–	2–3
Machine tools	Main	2.1	0.04	0.14	0.001	1–4
Punching and shearing machines	Main	28	0.10	–	0.001	1–2
	Crank pin	56	–	–	–	–
Rolling mills	Main	21	0.05	1.4	0.0015	1–1.5

Table 4.4 Dimensionless performance parameters for full journal bearings with side flow

$\left(\dfrac{l}{d}\right)$	ε	$\left(\dfrac{h_0}{C}\right)$	S	ϕ	$\left(\dfrac{r}{C}\right)f$	$\left(\dfrac{Q}{rCn_s l}\right)$	$\left(\dfrac{Q_s}{Q}\right)$	$\dfrac{p}{p_{max}}$
∞	0	1.0	∞	(70.92)	∞	π	0	–
	0.1	0.9	0.240	69.10	4.80	3.03	0	0.826
	0.2	0.8	0.123	67.26	2.57	2.83	0	0.814
	0.4	0.6	0.0626	61.94	1.52	2.26	0	0.764
	0.6	0.4	0.0389	54.31	1.20	1.56	0	0.667
	0.8	0.2	0.021	42.22	0.961	0.760	0	0.495
	0.9	0.1	0.0115	31.62	0.756	0.411	0	0.358
	0.97	0.03	–	–	–	–	0	–
	1.0	0	0	0	0	0	0	0
1	0	1.0	∞	(85)	∞	x	0	–
	0.1	0.9	1.33	79.5	26.4	3.37	0.150	0.540
	0.2	0.8	0.631	74.02	12.8	3.59	0.280	0.529
	0.4	0.6	0.264	63.10	5.79	3.99	0.497	0.484
	0.6	0.4	0.121	50.58	3.22	4.33	0.680	0.415
	0.8	0.2	0.0446	36.24	1.70	4.62	0.842	0.313
	0.9	0.1	0.0188	26.45	1.05	4.74	0.919	0.247
	0.97	0.03	0.00474	15.47	0.514	4.82	0.973	0.152
	1.0	0	0	0	0	0	1.0	–
$\dfrac{1}{2}$	0	1.0	∞	(88.5)	∞	π	0	–
	0.1	0.9	4.31	81.62	85.6	3.43	0.173	0.523
	0.2	0.8	2.03	74.94	40.9	3.72	0.318	0.506
	0.4	0.6	0.779	61.45	17.0	4.29	0.552	0.441
	0.6	0.4	0.319	48.14	8.10	4.85	0.730	0.365
	0.8	0.2	0.093	33.31	3.26	5.41	0.874	0.267
	0.9	0.1	0.031	23.66	1.60	5.69	0.939	0.206
	0.97	0.03	0.00609	13.75	0.610	5.88	0.980	0.126
	1.0	0	0	0	0	–	1.0	0

4.11 DESIGN OF JOURNAL BEARING

Journal bearing design involves a preliminary knowledge of following parameters:
1. Length to diameter ratio
2. Unit bearing pressure
3. Radial clearance
4. Minimum oil film thickness

4.11.1 Preliminary Design

4.11.1.1 Length to Diameter Ratio (l/d)

Increased length to diameter ratio (l/d) results in a long bearing which has more load carrying capacity as compared with short bearing.

Reduced length to diameter ratio (l/d) results in a short bearing which has greater side flow (Fig. 4.10) thereby improving heat dissipation.

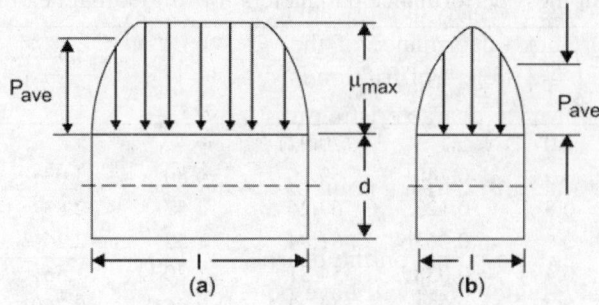

l/d = 2 for (a) and 1 for (b).

Fig. 4.10 l/d ratio *vs* bearing pressure

4.11.1.2 Unit Bearing Pressure

Load per unit of projected area on sliding contact journal bearing, is the 'unit bearing pressure' and depends upon bearing operating temperature, frequency of load and its nature, bearing material, etc.

4.11.1.3 Radial Clearance

Radial clearance is the spacing (radial) between journal (shaft) and bearing separated by lubricant film and this should be small to provide the necessary velocity gradient.

Practical value of radial clearance is 0.01 mm/mm of the journal radius.

i.e. $C = (0.001)r$.

4.11.1.4 Minimum Oil Film Thickness

Minimum oil film thickness below which metal to metal contact occurs and lubricant film may break is given by: $h_0 = (0.0002)r$, and is responsible for surface finish of journal (portion of shaft inside bearing).

Design of hydrodynamic journal bearing is done on the basis of:
- Maximum load carrying capacity
- Minimum frictional loss

4.11.2 Bearing Characteristic Number (ZN/p)

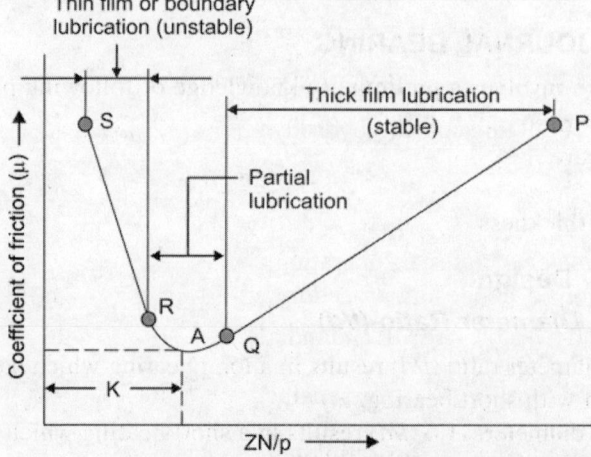

Fig. 4.11 Variation of μ with ZN/p

Factor ZN/p (Z—absolute viscosity of lubricant in kg/m-s, N—Speed of journal, p—Unit bearing pressure) helps to predict performance of the bearing (sliding contact). Variation of (ZN/p) operating values, with coefficient of friction is shown in Fig. 4.11.

Higher values of 'bearing characteristic number' results in hydrodynamic action and thin film lubrication.

With decreasing values of (ZN/p), a point of minimum coefficient of friction (μ) is reached at 'A'.

On further decrease, high points on the bearing surfaces begin to touch and coefficient of friction begins to rise. Region 'RS' will have boundary lubrication and hydrodynamic action does not apply.

Minimum coefficient of friction occurs at $(ZN/p) \leq 1$, which implies smoother surface and in turn greater load carrying capacity of bearing. In general, (ZN/p) value ranges between 2 and 5 (for practical purposes).

4.11.3 Sommerfield Number

$$S = \left(\frac{ZN}{p}\right)\left(\frac{d}{C}\right)^2$$

This number is useful in design of sleeve bearings (where d = Diameter of journal and C = Diametral clearance)

For practical purpose, Sommerfield number (S) is approximately taken as (14.3×10^6) with 'Z' being absolute viscosity (kg/m-s), N and p being speed of journal (rpm) and unit bearing pressure (N/mm^2) respectively.

4.11.4 Operating Pressure

$$p = \frac{ZN}{4.75 \times 10^6}\left(\frac{d}{C}\right)^2\left(\frac{l}{d+l}\right) \text{ N/mm}^2$$

(where 'l' being length and 'd' being diameter of bearing). At 'critical pressure $(P_{minimum})$', oil film breaks down and metal to metal contact begins.

4.11.5 Coefficient of Friction

$$\mu = \frac{33}{10^8}\left(\frac{ZN}{p}\right)\left(\frac{d}{C}\right) + K$$

where K is the factor for 'end leakage' correction = 0.002 (for l/d ratio between 0.75 and 2.8)

These values are for (sliding contact) hydrodynamic lubrication, as well as for self-contained full bearings.

4.11.6 Design Procedure

For given load, diameter of journal, operating speed of journal, operating temperature and lubricant supply system type, design procedure for sliding contact bearing is as follows:

1. Select l/d ratio from the table (ASME and Indian Standards)—design data for journal bearings.

2. Check bearing pressure $(p) = \dfrac{W}{ld}$ (for satisfactory value).

3. Assume a lubricant for specific operating temperature (t_o).

4. Find 'bearing characteristic number (ZN/p)' and compare it with corresponding table (ASME and Indian Standards), for maintaining fluid film operation.
5. Select clearance ratio 'C/d'.
6. Find coefficient of friction 'μ'.
7. Calculate heat generated and heat dissipated.
8. Thermal equilibrium needs to be maintained, so the heat dissipated should be at least equal to heat generated. If else, redesign the bearing with slightly different parameters.

WORKED EXAMPLES

[Refer Section 4.10, Design Data Handbook by Sadhu Singh]

1. Design a journal bearing to support a load of 7500 N while the shaft runs at 800 rpm using a hardened steel journal and bronze backed babbitt bearing. The clearance is 0.002 cm/cm diameter and the ambient temperature is 29 °C. Select a suitable oil and check whether artificial cooling is necessary.

SOLUTION:

$$\text{Given: Load } (F) = 7{,}500 \text{ N}$$

$$\text{Speed} = 800 \text{ rpm}$$

$$\frac{C}{d} = 0.002 \text{ (diametral clearance)}$$

$$\text{Ambient temperature} = 29\,°C$$

1. Let us select oil as "light transmission oil" and corresponding value (from Design Data Handbook by Sadhu Singh)

$$Z \approx 0.01 \text{ kg/m-s.}$$

2. Minimum bearing modulus for the given journal and bearing is 2.85.
Therefore, operating ZN/p is given as:

$$\frac{ZN}{p} = 3 \times 2.85 = 8.55$$

$$\therefore \qquad \frac{0.01 \times 800}{p} = 8.55 \text{ or } p = 0.936 \text{ MPa}$$

3. Required projected area $(l \times d) = \dfrac{F}{p} = \left(\dfrac{7{,}500}{0.936 \times 10^6} \right)$

Required projected area $\quad (A) = 0.00801 \text{ m}^2$

Let us take $\qquad\qquad l = 2d, \therefore 2d \times d = 0.00801$

On solving we get $\qquad d = 63.3 \text{ mm or } d = 64 \text{ mm}$

$\therefore \qquad\qquad\qquad l = 2d = 128 \text{ mm.}$

Actual bearing pressure

$$p = \frac{7{,}500}{0.064 \times 0.128} = 0.916 \text{ MPa}$$

4. Permissible pressure for film lubrication is given as

$$P = \frac{ZN}{4.75 \times 10^6} \left(\frac{d}{C} \right)^2 \left(\frac{l}{d+l} \right)$$

Putting values in

$$P = \frac{0.01 \times 800}{4.75 \times 10^6} \left(\frac{1}{0.001}\right)^2 \times \left(\frac{128}{64 + 128}\right) \text{ or } P = 1.22 \text{ MPa}$$

which is more than actual working pressure, therefore, the bearing will work under film lubrication.

5. Also friction factor $(f) = \dfrac{0.326}{10^6} \times \dfrac{ZN}{p} \times \dfrac{A}{d/C} + K_F$

$$f = \frac{0.326}{10^6} \times \frac{0.01 \times 800}{0.916} \times 1{,}000 + 0.02$$

$$f = 0.00485$$

6. Heat generation

$$H_g = tF_v \text{ (wall)}$$

also

$$V = \frac{\pi \times 64 \times 800}{1{,}000 \times 60} = 2.67 \text{ m/sec} \left(\frac{\pi dN}{60}\right)$$

∴

$$H_g = 0.00485 \times 7{,}500 \times 2.67$$

$$H_g = 97.5 \text{ W}$$

7. Heat dissipated

$$H_d = K_d \times dl \, (t_b - t_a)$$

For bearing, oil temperature is taken as about 80 °C

∴

$$t_b - t_a = \frac{1}{2}(80 - 29) = 25.5$$

Let

$$K_d = 390 \text{ for still air}$$

∴

$$H_d = 390 \times 0.064 \times 0.128 \times 25.5$$

⇒

$$H_d = 81.46 \text{ W}$$

Since

$$H_d > H_g$$

So artificial cooling is not required.

2. Design a journal bearing to support a load of 4 kN at 500 rpm, using a hardened steel journal and a bronze backed babbit bearing. Take room temperature as 20 °C and operating oil temperature as 70 °C.

SOLUTION:

Given: Load	(f)	$= 4 \times 10^3$ N
Speed	(N)	$= 500$ rpm
Room temperature	(t_a)	$= 70$ °C
Operating oil temperature	(t_b)	$= 70$ °C

Step 1

Let us assume the oil as light transmission oil for which $Z \approx 0.01$ kg/m-s (from Design Data Handbook by Sadhu Singh)

Step 2

From data book, minimum bearing modulus for the given journal and bearing is 2.85, therefore, operating (ZN/p) is given as:

$$\frac{ZN}{p} = 3 \times 2.85 = 8.55$$

$$\therefore \qquad p = \frac{ZN}{8.55} = \frac{0.01 \times 500}{8.55} = 0.585 \text{ MPa}$$

Step 3

Required projected area $(l \times d) = F/p$

or
$$l \times d = \frac{4 \times 10^3}{0.585 \times 10^6} = 6.84 \times 10^{-3} \text{ m}^2$$

Let us take $\qquad l = 2d$

$\therefore \qquad 2d \times d = 6.84 \times 10^{-3}$ or $d^2 = 3.42 \times 10^{-3}$

or $\qquad d = 0.0585$ m, or $d = 58.5$ mm.

and length of bearing $(l) = 2d = 0.117$ m

or $\qquad l = 117$ mm.

$\therefore$ Actual bearing pressure $= \dfrac{4 \times 10^3}{58.5 \times 117 \times 10^{-6}}$

Bearing pressure $\quad (p) = 0.584 \times 10^6 \text{ N/mm}^2$.

Step 4

Permissible pressure for film lubricant

$$P = \frac{ZN}{4.75 \times 10^6} \left(\frac{d}{C}\right)^2 \left(\frac{l}{d+l}\right)$$

Let's take $\qquad C/d = 0.001$

$\therefore \qquad P = \dfrac{0.01 \times 500}{4.75 \times 10^6} \times 10^6 \times \left(\dfrac{117}{58.5 + 117}\right)$

$$P = 0.7017 \text{ MPa}$$

which is more than actual working pressure, therefore, the bearing will work under film lubrication.

Step 5

Now friction factor $\quad (f) = \dfrac{0.326}{10^6} \times \dfrac{ZN}{p} \times \dfrac{d}{C} \times K_F$

$$= \frac{0.326}{10^6} \times \frac{0.01 \times 500}{0.584} \times 1,000 + 0.002$$

or $\qquad f = 4.79 \times 10^{-3}$

Step 6

Heat generated $(H_g) = f F V$ watt

where $\qquad V = \dfrac{\pi d N}{60} = \pi \times \dfrac{58.5}{1,000} \times \dfrac{500}{60}$

or $\qquad V = 1.531$ m/sec

$\therefore \qquad H_g = 4.79 \times 10^{-3} \times 4 \times 10^3 \times 1.531$

$\qquad H_g = 29.34$ watt $\qquad\qquad\qquad$ ...(1)

Step 7

Heat dissipated $(H_d) = K_d \, (dl)(t_b - t_a)$

from data book, $K_d = 390$ for still air and

$\qquad (t_b - t_a) =$ Average temperature difference

$$= \frac{1}{2}(70 - 20) = 25°C$$

$\therefore \qquad H_d = 390 \times (58.5 \times 117 \times 10^{-6}) \times 25$

$\therefore \qquad H_d = 66.7$ watt. $\qquad\qquad\qquad$ ...(2)

Comparing (1) and (2) we get $H_d > H_g$

So design is safe.

3. Design a journal bearing for a centrifugal pump for the following data.

$\qquad$ Load on journal = 15,000 N

$\qquad$ Speed of journal = 900 rpm

$\qquad$ Type of oil = SAE10

$\qquad$ Operating temperature = 55 °C

$\qquad$ Ambient temperature of oil = 15.5 °C

$\qquad$ Maximum bearing pressure for pump = 1.5 N/mm^2

Also calculate the mass of lubricating oil required for artificial cooling, if rise of temperature of oil be limited to 15 °C. Heat dissipation coefficient is 1,232 W/m^2/°C.

SOLUTION:

$\qquad$ Given: Load on journal (W) $\qquad$ = 15,000 N

$\qquad$ Speed of journal (N) $\qquad\qquad$ = 900 rpm

$\qquad$ Type of oil $\qquad\qquad\qquad\qquad$ = SAE 10

$\qquad$ Operating temperature $\qquad\quad$ = 55 °C

$\qquad$ Ambient temperature of oil $\quad$ = 15.5 °C

$\qquad$ Maximum bearing pressure (p) = 1.5 N/mm^2 = 1.5 MPa

$\qquad$ Rise of temperature of oil $\quad\;$ = 15 °C

$\qquad$ Heat dissipation coefficient $\;$ = 1,232 W/m^2/°C

$$\text{Bearing pressure } (p) = \frac{W}{\text{Projected area of bearing}}$$

or $\qquad\qquad\qquad p = W/ld$

From (Design Data Handbook by Sadhu Singh) data table, l/d ratio for pumps varies from 1 to 2.

$\qquad$ Let us take $l = 1.5d$

$\qquad\qquad\qquad l =$ Length of bearing

$\qquad\qquad\qquad d =$ Diameter of bearing

$\therefore \; 1.5 \times 10^6 = \left(\dfrac{15,000}{1.5d \times d} \right)$

or $\qquad\qquad d = 0.0816$ m, or $d = 81.6$ mm

or $\qquad\quad d = 82$ mm and $l = 1.5 \times 82 \approx 125$ mm (say)

Actual bearing pressure $(p) = \dfrac{W}{ld} = \dfrac{15,000}{82 \times 124 \times 10^{-6}}$

or $\qquad p = 1.475 \times 10^{-6}$ Pa, $p = 1.475$ MPa.

Viscosity of SAE10 at operating temperature of 55 °C is $\eta = 14$ CP (from Design Data Handbook).

Sommerfield number $(K) = \left(\dfrac{\eta \times N}{p}\right)\left(\dfrac{d}{C}\right)^2$

$\qquad$ assuming $\left(\dfrac{C}{d}\right) = 0.001$

$\therefore \qquad\qquad K = \left(\dfrac{14 \times 900}{1.47 \times 10^6 \times 60}\right)\left(\dfrac{1}{0.01}\right)^2$

$\qquad\qquad\qquad = 142$

From MC Kel's equations for coefficient of friction for light loaded bearing and K is about 150

$$\mu = 2\pi^2\left(\dfrac{ZN}{p}\right)\left(\dfrac{d}{C}\right) + K$$

K is a constant which accounts for end leakage for $0.75 < \dfrac{l}{d} < 2.8$.

So value of $K \approx 0.002$

$\therefore$ Coefficient of friction $(\mu) = 2\pi^2\left[\dfrac{14 \times 10^{-3} \times 900}{1.47 \times 10^6 \times 60}\right]\left[\dfrac{1}{0.001}\right] + 0.002$.

$\Rightarrow \qquad\qquad\qquad \mu = 0.00482$

Speed of journal $(V) = \pi dN = \pi \times 82 \times 10^{-3} \times \dfrac{900}{60}$

or $V \qquad\qquad = 3.9$ m/sec.

Heat generated $(H_g) = \mu pldV = 0.00482 \times 1.47 \times 10^6 \times 124 \times 82 \times 10^{-6} \times 3.9$

$\qquad\qquad H_g = 280$ W

This heat is removed by lubricating oil.

Let m = Mass of the lubricating oil required.

Heat generated = Heat removed by oil

$\qquad\qquad 280 = mC \times \Delta T$

$\qquad\qquad 280 = m \times 1232 \times 15$

or $\qquad\qquad m = 0.0152$ kg/sec or $m = 54.73$ kg/hour

PREVIOUS YEAR UNIVERSITY QUESTIONS

$\qquad\qquad$ *[Refer section 4.10, Design Data Hand Book by Sadhu Singh]*

1. Design a journal bearing for a centrifugal pump from the following data.

$\qquad$ Load on the journal $\qquad\qquad$ = 20 kN

$\qquad$ Speed of the journal $\qquad\qquad$ = 1,000 rpm

$\qquad$ Absolute viscosity of oil at 55 °C = 0.017 kg/m-s

$\qquad$ Ambient temperature of oil $\qquad$ = 16 °C

$\qquad$ Maximum bearing pressure for the pump = 1.5 N/mm². $\qquad\qquad$ (UPTU 2008)

SOLUTION:

Given: $W = 20$ kN $= 20,000$ N

$N = 1,000$ rpm

$T_0 = 55\,°C$, $Z = 0.017$ kg/m-s, $t_a = 16\,°C$, $p = 1.5$ N/mm^2

Length of the journal

$l = 1.6d = 1.6 \times 100 = 160$ mm

(assuming diameter of journal (d) as 100 mm)

Bearing pressure $(p) = \dfrac{W}{ld} = \dfrac{20,000}{160 \times 100} = 1.25$ N/mm^2

as this is less than given bearing pressure (1.5 N/mm^2)
So values of l and d are safe.

$$\frac{ZN}{p} = \frac{0.017 \times 1,000}{1.25} = 13.6$$

From Data Book, operating value of $\dfrac{ZN}{p} = 28$

So

$$3K = \frac{ZN}{p}$$

∴ Bearing modulus at the maximum point of friction

$$K = \frac{1}{3}\left(\frac{ZN}{p}\right) = \frac{1}{3} \times 28 = 9.33$$

Since the calculated value of bearing characteristic number, $\dfrac{ZN}{p} = 13.6$ is more than 9.33,

therefore, the bearing will operate under hydrodynamic conditions. From Data Book, for
centrifugal pumps, the clearance ratio (C/d) = 0.0013.

Coefficient of friction $(\mu) = \dfrac{33}{10^8}\left(\dfrac{ZN}{p}\right)\left(\dfrac{d}{C}\right) + K$

$$= \frac{33}{10^8} \times 13.6 \times \frac{1}{0.0013} + 0.02$$

$$= 0.00545$$

Heat generated $(Q_g) = \mu WV = 0.00545 \times 20,000 \left(\dfrac{\pi dN}{60}\right)$ W

$$= 0.0545 \times 20,000 \left(\frac{\pi \times 0.1 \times 1,000}{60}\right)$$

$$= 570.72$$ W

2. Design a journal bearing for a centrifugal pump. The shaft diameter is 150 mm and length to
diameter ratio is 1.6 and the load on the bearing is 40 kN. The speed of shaft is 1,500 rpm.

(UPTU 2010)

SOLUTION:

Given: $\dfrac{l}{d} = 1.6$, $d = 150$ mm, $W = 40$ kN, $N = 1,500$ rpm

$l = 1.6 \times d = 1.6 \times 150 = 240$ mm

or
$$p = \frac{W}{ld} = \frac{40 \times 10^3}{240 \times 150} = 1.11 \text{ N/mm}^2$$

$$\frac{ZN}{p} = \frac{0.017 \times 1,500}{1.11} = 22.972.$$

Clearance ratio $\left(\dfrac{C}{d}\right) = 0.013$

So coefficient of friction $(\mu) = \dfrac{33}{10^8}\left(\dfrac{ZN}{p}\right)\left(\dfrac{d}{C}\right) + K$

$$= \left(\frac{33}{10^8}\right)(22.972)\left(\frac{1}{0.013}\right) + 0.002 = 0.0078.$$

So heat generated $(H_g) = \mu WV = 0.078 \times 40 \times 10^3 \times \dfrac{\pi dN}{60 \times 10^3}$

$$= 0.078 \times 40 \times 10^3 \times \frac{\pi \times 150 \times 1,500}{60 \times 10^3}$$

$$= 36,738 \text{ W}$$

3. A full journal bearing of diameter 80 mm and length 120 mm is to support a load of 20 kN at the shaft speed of 1,500 rpm. The bearing temperature is to be limited to 75 °C and the ambient room temperature is 38 °C. The viscosity oil used is 0.0088 kg/m-s at 115 °C. Check, if artificial cooling is required and find the amount of artificial heating.   (UPTU 2011)

SOLUTION: Given: $d = 80$ mm, $l = 120$ mm.

Bearing pressure $(p) = \dfrac{W}{A} = \dfrac{W}{ld} = \dfrac{20,000}{120 \times 80} = 2.08 \text{N/mm}^2$

Coefficient of friction $(\mu) = \dfrac{33}{10^8}\left(\dfrac{ZN}{p}\right)\left(\dfrac{d}{C}\right) + 0.002$

$$= \frac{33}{10^8}\left(\frac{0.0088N}{2.08}\right)\left(\frac{80}{0.1}\right) + 0.02$$

$$= \frac{111.69}{10^8} N + 0.002$$

Heat generated $(H_g) = \mu WV$

$$= \mu W\left(\frac{\pi dN}{60}\right) \text{ J/sec}$$

$$= \left(\frac{111.69N}{10^6} + 0.002\right) \times 20,000\left(\frac{\pi \times 0.08N}{60}\right)$$

$$= \frac{9,356.919}{10^8} N^2 + 0.168 N$$

$$= \frac{9,356.919 \times 1,500}{10^8} + 0.168 \times 1,500 = 252.14 \text{ J/sec.}$$

also $t_b = 75\,°\text{C}$ and $t_a = 38\,°\text{C}$ (t_b—bearing surface temperature)

So $(t_b - t_a) = \frac{1}{2}(t_b - t_a) = \frac{1}{2}(75 - 38) = 18.5\,°C.$

From Data Book (at 18.5 °C)

$$C = 280 \text{ W/m}^2/°C.$$

Heat dissipated
$$(Q_d) = CA(t_b - t_a)$$
$$= 280 \times 0.120 \times 0.08 \times (75 - 38)$$
$$= 99.46 \text{ J/sec.}$$

So amount of artificial heat required
$$= 252.14 - 99.46$$
$$= 152.68 \text{ J/sec.}$$

4. Design a journal bearing to support a load of 5 kN at 1,000 rpm using a hardened steel journal and bronze backed babbit bearing. The bearing is lubricated by oil rings. Assume the room temperature as 25 °C and the oil temperature as 77 °C.

SOLUTION:

Given: $W = 5$ kN $= 5,000$ N, $N = 1,000$ rpm, $t_a = 25°C$, $t_o = 77°C$.

Assume diameter of journal (d) as 100 mm.

From Design Data Handbook, ratio of l/d for centrifugal pump varies from 1 to 2.

Let us take $l/d = 1.6$

So $l = 1.6d = 1.6 \times 100 = 160$ mm.

Bearing pressure $(p) = \dfrac{W}{ld} = \dfrac{5,000}{160 \times 100} = 0.3125 \text{ N/mm}^2$

$$\frac{ZN}{p} = \frac{0.017 \times 1,000}{0.3125} = 54.4$$

From Design Data Handbook, $\dfrac{ZN}{p} = 28.$

The minimum value of the bearing modulus at which the oil film will break is given by:

$$3K = ZN/p$$

Bearing modulus at the minimum point of friction $(K) = \dfrac{1}{3}\left(\dfrac{ZN}{p}\right) = \dfrac{1}{3} \times 28 = 9.33$

Since the calculated value of bearing characteristic number $\left(\dfrac{ZN}{p}\right) = 54.4$ is more than 9.33, therefore, the bearing will operate under hydrodynamic conditions.

From Data Book, the clearance ratio $(C/d) = 0.0013$ (it is selected as coefficient of friction)

$$\mu = \frac{33}{10^8}\left(\frac{ZN}{p}\right)\left(\frac{d}{C}\right) + K$$

(taking $K \approx 0.002$) → (from Design Data Handbook)

So
$$\mu = \frac{33}{10^8} \times 54.4 \times \frac{1}{0.0013} + 0.002$$
$$= 0.0158$$

So heat dissipated

$$Q_d = CA(t_b - t_a) = Cld(t_b - t_a)\text{W}$$

$$(t_b - t_a) = \frac{1}{2}(t_o - t_a) = \frac{1}{2}(77 - 25) = 26\ ^\circ\text{C}$$

$$Q_d = 1{,}232 \times 0.16 \times 0.1 \times 26 = 512.51\ \text{W}$$

4.12 HYDROSTATIC THRUST BEARING

Refer Section 4.2 (Hydrostatic Thrust Bearings). These find application in vertical steam turbines, motor and pumps, and hydraulic turbines, and support axial/longitudinal load along shaft axis.

These are classified as:

1. Pivot/step bearings: Support loaded shaft in vertical positions and called as pivot or footstep bearing (refer Section 4.2, Hydrostatic Footstep Bearing).
2. Collar bearing: These include collar supporting the shaft and are used in horizontal as well as vertical shafts (refer Section 4.2, Collar Bearing).

4.13 THRUST BEARING: PIVOT AND COLLAR BEARINGS

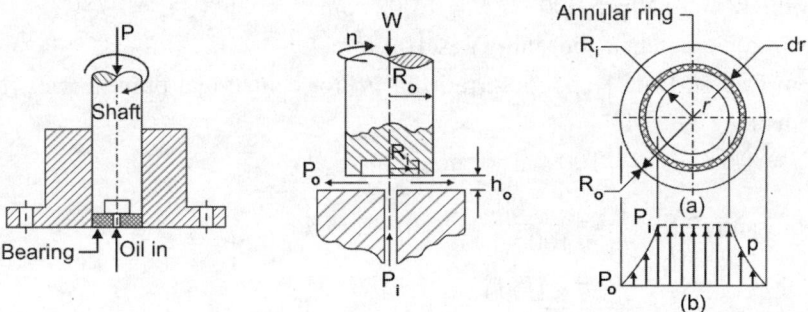

Pressure distribution in hydrostatic bearing

Fig. 4.12 Pivot bearing

4.13.1 Pivot Bearing

Considering design criteria of pivot bearing. Wear will be maximum at the outer radius and minimum at the center of the shaft because wear being proportional to rubbing velocity, which in turn is proportional to radius of shaft.

This wear results in overheating and failure of the lubricant film.

This problem is countered by:

1. Supporting shaft on a pile of alternate discs of different materials like steel and bronze.
2. Counterboring shaft end to a short depth to reduce endwear.

Base of bearing surface can be of cast iron, bronze or steel with a lining of babbit and base is either pinned or keyed to prevent rotation. Bronze bushing takes the lateral wear.

If W = Bearing load

                  R = Radius of shaft/bearing

                  A = Bearing surface area

                  p = Bearing pressure/unit area

                  μ = Coefficient of friction

                  N = Shaft speed in rpm

then $p = \dfrac{W}{A} = \dfrac{W}{\pi R^2}$

Frictional torque $(\tau) = \left(\dfrac{2}{3}\right)\mu WR$

Power lost in friction $(P) = \dfrac{2\pi N\tau}{60}$ watt

While using a counterbored shaft: $p = \dfrac{W}{\pi(R^2 - r^2)}$, with r = Counterbore radius

and $\tau = \left(\dfrac{2}{3}\right)\mu W\left(\dfrac{R^3 - r^3}{R^2 - r^2}\right)$

WORKED EXAMPLE

[Refer Section 4.10, Design Data Handbook by Sadhu Singh]

1. A shaft of 150 mm diameter is supported in a footstep bearing which is counterbored at the end with a hole diameter of 50 mm. The speed of the shaft is 100 rpm and the allowable bearing pressure is 0.8 N/mm^2. Determine: (i) Load which can be supported, (ii) power lost in friction and (iii) the heat generated in bearing. Take $\mu = 0.015$.

SOLUTION: Load on the shaft:

$$p = \dfrac{W}{\pi(R^2 - r^2)}$$

where
p = Pressure over the bearing surface
W = Load on the shaft
R = Outer radius of the shaft
r = Outer radius of the hole

$$0.8 = \dfrac{W}{\pi(75^2 - 25^2)} = \dfrac{W}{15710}$$

This gives $W = 0.8 \times 15{,}710 = 12{,}568$ N.

Power lost in friction

Frictional torque

$$(\tau) = \left(\dfrac{2}{3}\right)\mu W\left[\dfrac{R^3 - r^3}{R^2 - r^2}\right]$$

$$= \left(\dfrac{2}{3}\right) \times 0.015 \times 12{,}568\left[\dfrac{75^3 - 25^3}{75^2 - 25^2}\right]$$

$$= 125.68 \times 81.25 = 10{,}212 \text{ N-mm}$$

$$= 10.212 \text{ N-m}.$$

Power lost in friction $= \dfrac{2\pi N\tau}{60} = \dfrac{2\pi \times 100 \times 10.212}{60} = 107$ W

or $= 0.107$ kW

Heat generated in bearing

Power lost in friction is converted into the heat.

Therefore, heat generated $= 0.107 \times 60 = 6.42$ kJ/min.

4.13.2 Collar Bearing

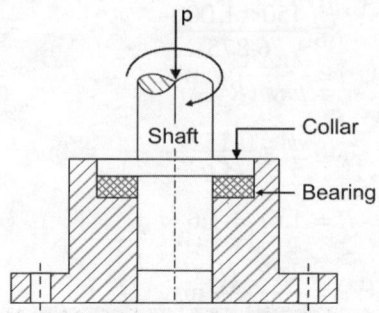

Fig. 4.13 Collar bearing

Bearing pressure gets distributed over a number of collars. Total axial load on bearing is given as.

$$W = pn\pi (R^2 - r^2)$$

or

$$p = \frac{W}{n\pi (R^2 - r^2)}$$

where n = Number of collars, r = Inner collar radius,
R = Outer collar radius and W = Load transmitted along axis.

Design Parameters

- R/r lies between 1.4 and 1.8
- Collar width: $r/3$
- Clearance/spacing between collars: $1.5r$
- Allowable bearing pressure: Product of pressure and rubbing/sliding
  Velocity: $PV \leq 700$.
  ('P' in 'kPa' and 'V' in m/sec)
  here V lies between 0.25 and 10 m/sec.
  For $V > 1$ m/sec, $P \leq 0.7$ MPa (around 0.35–0.42 N/mm^2)
  For $V < 1$ m/sec, $P \approx 14$ MPa
  and for intermittent service, $P \approx 10.5$ MPa.

WORKED EXAMPLE

[Refer Section 4.10, Design Data Handbook by Sadhu Singh]

1. A propeller shaft of 450 kW power of massive oil engine is provided with a multicollar thrust bearing.

The shaft diameter is 0.15 m and runs at 220 rpm. The propeller has a pitch of 2.50 m. Assuming a slip of 25%, determine the main dimensions of bearing.

SOLUTION: A pitch of 2.50 m results in a forward movement of 2.5 m for each revolution of the engine.

With the given slip of 25%, the actual distance moved

$$= 2.50 - \frac{25}{100} \times 2.50$$

$$= 1.875 \text{ m for each revolution of engine}$$

$$= 1.875 \times \frac{220}{60} = 6.875 \text{ m/sec}$$

Thrust on the shaft (W) = Power × Speed

$$= \frac{450 \times 1,000}{6.875} = 65.50 \text{ kN}$$

Further $\qquad\qquad W = pn\pi \, (R^2 - r^2)$

Since $\qquad\qquad r = \frac{d}{2} = \frac{0.15}{2} = 0.075 \text{ m}$

Take $\qquad\qquad R = 1.6r = 1.6 \times 0.075 = 0.12 \text{ m}$

Mean rubbing/sliding velocity $(V) = \frac{2\pi R_m N}{60}$ m/sec

$\Rightarrow \qquad\qquad R_m = \frac{0.12 + 0.075}{2} = 0.0975 \text{ m}$

$$V = \frac{2\pi \times 0.0975 \times 220}{60} = 2.24 \text{ m/sec}$$

Since $V > 1$ m/sec, therefore, allowable bearing pressure should be less than 0.7 MPa. Take $p = 0.5$ MPa.

$\therefore \qquad\qquad 65.50 \times 10^3 = 0.5 \times 10^6 \times n \times \pi \, (0.12^2 - 0.075^2)$

This gives $n = 4.76$ say 5 collars

Total torsional moment $\left(T_f\right) = \mu W \frac{(R + r)}{2}$

$$= 0.03 \times 65.50 \times 1000 \times \frac{0.195}{2} \text{ (assuming } \mu = 0.03)$$

$$= 191.5 \text{ Nm.}$$

So loss of power $= T_f \times W$

$$= \frac{191.5 \times \pi \times 220}{30} = 4410 \text{ W}$$

$$= 4.41 \text{ kW}$$

KEY TERMS

- Thick film lubrication
- Hydrodynamic bearing
- Hydrostatic bearing
- Thin film lubrication
- Bearing characteristic number *vs* coefficient of friction graph
- Design parameters (sliding contact bearings): Eccentricity, radial clearance, minimum film thickness (h_0), eccentricity ratio (ε), Sommerfield number, unit bearing pressure, flow variable (FV), coefficient of flow variable (CFV).
  - Bearing material
  - *l*/*d* ratio, radial clearance, minimum oil film thickness, maximum oil film temperature.
  - Heat generated in bearing.

SUMMARY

- When two surfaces of bearing in relative motion are completely separated by a fluid film and surface friction has no role, it is called 'thick film lubrication' which is classified as "hydrodynamic and hydrostatic".
- Hydrodynamic bearing involves creation of load-supporting fluid film, by shape and relative motion of sliding surface. The only requirement is sufficient and continuous supply and finds application in engine and centrifugal pump.

  Also bearing without lubricant, is called 'zero film bearing'.
- Journal bearing is a sliding contact bearing working on hydrodynamic lubrication and support the load in radial direction. It has two types: Full journal and partial journal bearings.
- Hydrostatic bearing involves, load-supporting fluid film separating two surfaces being created by external source like pump. Supplying sufficient fluid under pressure and find application in centrifuges, ball mills and vertical turbo generator.
- Thin film lubrication comprise of a relatively thin lubricant film along with partial metal to metal contact. In this lubrication under excessive load, less oil supply, low speed, misalignment and boundary lubrication will occur.
- Bearing characteristic number = $\dfrac{ZN}{p}$ (determines performance of bearing)

  with 'Z' being absolute viscosity, 'N' being speed of shaft in rpm and 'p' being unit bearing pressure.
- Radial clearance $(C) = R - r$. (R = Bearing radius and r = Shaft radius)
- Minimum oil film thickness $(h_0) = R - (r + e)$
- Eccentricity ratio $(\varepsilon) = \dfrac{e}{C} = 1 - h_0/C$

  $\dfrac{h_0}{C}$ = Minimum thickness variable
- Sommerfield number $(S) = \left(\dfrac{r}{C}\right)^2 \dfrac{ZN}{p}$
- Flow Variable $(FV) = \dfrac{Q}{NrCl}$ (r is shaft radius and l is bearing length)
- Coefficient of flow variable $(CFV) = \left(\dfrac{r}{C}\right)\mu$
- Temperature rise $(\Delta t) = \dfrac{8.3 p (CFV)}{(FV)}$ (p is unit bearing pressure)
- Radial clearance $(C) = \dfrac{r}{1,000}$ (empirically)
- Minimum oil film thickness

  $$h_{0(min)} = \dfrac{r}{5,000}$$
- Maximum oil film temperature < 120 °C, to prevent oxidation of oil.
- Bearing material should have high compressive strength, endurance strength (to avoid pitting), conformability and embedability.
- Babbitt is a most popular bearing (sliding contact) material. Due to its silvery appearance, it is called 'white metal'. Two varieties are lead based and tin based.

- Heat generated in bearing or power loss = μWV
  (where W is radial load and V is rubbing velocity)

- Heat dissipated $(H_d) = \dfrac{(\Delta t + 18)^2 \times A}{K}$ watt

  where Δt = Bearing temperature, $A = l \times d$

  K = 0.484 (for unventilated bearing)

  = 0.273 (for well-ventilated bearing)

REVIEW QUESTIONS

Short Answers

1. Define lubrication along with objectives.
2. Name any two liquid, semi-solid and solid lubricants.
3. What is thick film and zero film bearings?
4. What is hydrodynamic lubrication?
5. Give two applications of hydrodynamic and hydrostatic bearings.
6. What is SAE?

Long Answers

1. Explain the procedure followed in designing a journal bearing.
2. List and define the important physical characteristics of a good bearing material.
3. Explain wedge film and squeeze film lubrication.
4. What are journal bearings? Give classification of three bearings.
5. State and explain the types of bearing failure.
6. Explain full journal and partial journal bearings.

Numericals

[Refer Section 4.10, Design Data Handbook by Sadhu Singh]

1. Following data is given for a full hydrodynamic bearing:

   Radial load = 25 kN

   Journal speed = 900 rpm

   Unit bearing pressure = 2.5 MPa

   (l/d) ratio = 1

   Viscosity of lubricant = 20 CP

   Class of fit = H7e7

   Calculate:

   i. Dimensions of the bearing
   ii. Minimum film thickness
   iii. Requirement of oil flow.

2. The load on a journal bearing is 3 kN, diameter 50 mm, length 75 mm, speed 1,600 rpm, diametral clearance 0.001 mm, ambient temperature 15.5 °C. Oil SAE10 is used and the film temperature is 60 °C. Determine the heat generated and the heat dissipated. Absolute viscosity of SAE10 at 60 °C = 0.014 kg/m-s.

3. Design a full hydrodynamic journal bearing with following specification for machine tool application.

Journal diameter	= 75 mm
Radial load	= 10 kN
Journal speed	= 1,440 rpm
Minimum oil film thickness	= 22.5 microns
Inlet temperature	= 40 °C
Bearing material	= Babbitt

Determine the length of the bearing and select suitable oil for this application.

4. A 360° hydrodynamic bearing operates under the following conditions:

Radial load	= 50 kN
Journal diameter	= 150 mm
Bearing length	= 150 mm
Radial clearance	= 0.15 mm
Viscosity of lubricant	= 8 CP

What is the minimum speed of operation for the journal to work under hydrodynamic conditions?

5

Rolling Contact Bearings

5.1 INTRODUCTION

Bearings with 'rolling contact (point contact)' between its element instead of 'sliding contact' (surface or line contact) are called rolling contact bearings.

Rolling contact bearings have very less friction as compared to sliding contact at the time of start, but once the condition of full hydrodynamic lubrication is attained in sliding contact bearings, friction is less.

These are classified as: Ball bearings and roller bearings.

Ball bearings mainly find application for light loads, whereas roller bearings find application for heavier loads.

Bearings in general can be defined as mechanical element that permits relative motion between two parts, such as the shaft and the housing; with minimum friction. Main functions of bearings include: Ensuring free rotation of the shaft or the axle with minimum friction, supporting the shaft or the axle and holding it in correct position; taking up the forces that act on the shaft or the axle and transmitting them to the frame or foundation.

Rolling contact bearings are also called antifriction bearings and comprise rolling elements like balls or rollers, introduced between the surfaces that are in relative motion. Friction involved is rolling friction.

These find application in:

1. Gear boxes, 2. Automobiles axles, 3. Machine tool spindles, 4. Electric motors, 5. Crane Hook and hoisting drum.

Balls and inner/outer races of roller contact bearings are made of high carbon chromium steel (SAE 52100 or AISI 5210). Balls and races are thorough hardened to obtain a minimum hardness of 58 Rockwell C. Rollers are made of case hardened steels (AISI 3310, 4620 or 8620).

A rolling contact bearing consists of four parts:

- Inner and outer races
- Rolling elements—ball, cylindrical roller, etc.
- Cage (holding rolling elements together and spacing them evenly around the periphery of the shaft.)

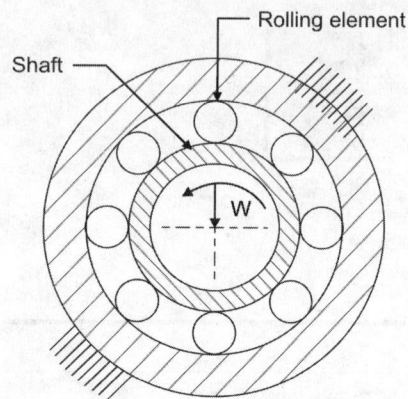

Fig. 5.1 Rolling contact bearing

On the basis of types of rolling element used, bearings are classified as ball bearings, roller bearings, etc.

Rollers are case carburized to obtain a surface hardness of 58 Rockwell C. Balls are thorough hardened while rollers are case hardened.

5.2 TYPES OF BALL BEARINGS AND ROLLER BEARINGS

Depending upon types of rolling element used, bearings are classified as:

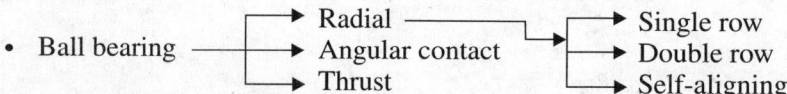

- Ball bearing
  - Radial — Single row
  - Angular contact — Double row
  - Thrust — Self-aligning

- Cylindrical roller bearing
- Taper roller bearing
- Needle bearing

Depending upon direction of load, rolling contact bearings are classified as:

- Radial bearing
- Thrust bearing

On this basis types of rolling contact bearings, frequently used are:

a. Deep groove ball bearing
b. Cylindrical roller bearing
c. Angular contact bearing
d. Self-aligning bearing
e. Taper roller bearing
f. Thrust ball bearing

Angular contact bearings are used when loads are combined radial and high axial thrust and when accurate shaft location is required whereas thrust bearings are used exclusively for carrying axial thrust loads and speed below 200 rpm.

5.3 ADVANTAGES AND DISADVANTAGES

Deep Groove Ball Bearing

Advantages

1. Has high load carrying capacity owing to large size of the balls.
2. Takes load in radial as well as axial direction.
3. Frictional loss and resultant temperature rise is less in this type of bearing. Deep groove ball bearings give excellent performance in high speed applications as maximum permissible speed of shaft depends upon temperature rise of the bearing.
4. Generates less noise.
5. Available within wide range of bore diameters (few mm to 400 mm).

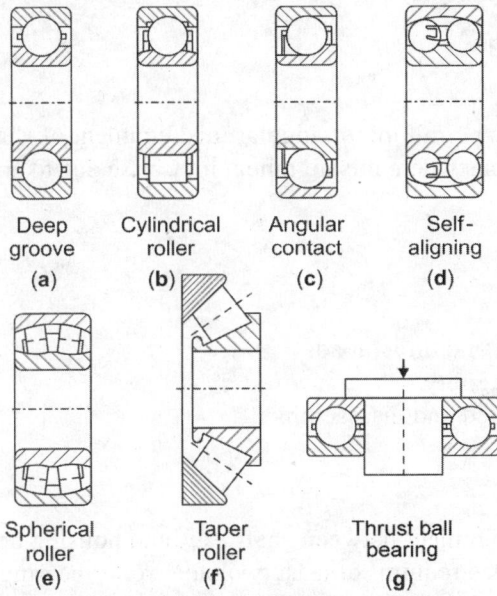

Deep groove **(a)**   Cylindrical roller **(b)**   Angular contact **(c)**   Self-aligning **(d)**

Spherical roller **(e)**   Taper roller **(f)**   Thrust ball bearing **(g)**

Fig. 5.2 Types of rolling contact bearing

Disadvantages

1. Self-alignment is not present and alignment between axis of shaft and **housing bore** is required for accuracy.
2. Have poor rigidity, due to point of contact and is unsuitable for applications like machine tool spindles requiring rigidity as main criteria.

Cylindrical Roller Bearing

Advantages

1. Load carrying capacity is quite high owing to line contact between rollers.
2. More rigid than ball bearing.
3. Coefficient of friction and thereby frictional loss is low in high speed applications.

Disadvantages

1. Axial or thrust loads cannot be supported.
2. No self-alignment and needs precise alignment, between shaft axis and bore housing.
3. Generates more noise, owing to line of contact.

Angular Contact Bearing

Advantages

1. Can support both radial and axial/thrust loads.
2. Load carrying capacity is more than that of deep groove ball bearing on account of provision of inserting more number of balls than deep groove ball bearing by cutting away one side of groove in outer race.

Disadvantages

1. These require initial pre-loading.
2. These bearing must be mounted without axial movement.
3. More than one bearing is required to take thrust load in either direction.

Self-aligning Bearings

Advantages

1. Self-aligning bearings permit minor angular misalignment of shaft relative to the housing.
2. Suitable for applications where misalignment may arise due to errors in mounting or due to deflection of the shaft.

Taper Roller Bearing

Advantages

1. Can take heavy radial and thrust loads.
2. Bearings have more rigidity.
3. Can be easily assembled and disassembled.

Disadvantages

1. Costly
2. Cannot tolerate misalignment between shaft axes and housing bore.
3. Balancing of axial force requires at least two taper roller bearings.

Thrust Ball Bearings

Advantages

1. Use of large number of balls results in high thrust load carrying capacity in smaller space, e.g. in worm gear boxes and crane hooks.

Disadvantages

1. Bearings cannot take radial loads.
2. Not self-aligning and cannot tolerate misalignment.
3. Operate better on a vertical shaft as compared to horizontal shafts.
4. Require continuous pressure applied by springs to hold rings (shaft ring and housing ring) together.

5.4 THRUST BALL BEARING

Ball bearings used to mainly support axial loads only are thrust ball bearings. These comprise a row of balls running between two rings (shaft ring and housing ring). Thrust ball bearings cannot carry radial load and carries heavy thrust loads in only one direction, using large number of balls resulting in high thrust load carrying capacity in smaller space.

Advantages and disadvantages (refer Section 5.3, Thrust Ball Bearings).

Applications of thrust ball bearings include application of heavy thrust loads, e.g. in worm gear boxes and crane hooks. Thrust ball bearings give satisfactory performances at low and medium speeds, but at high speeds give poor service, owing to generation of centrifugal forces and gyroscopic couple on balls.

5.5 TYPES OF ROLLER BEARING

Roller bearing can be specially categorized on the basis of type of roller.
- Cylindrical roller
- Spherical roller
- Taper roller

Roller bearings have higher load carrying capacity as compared to ball bearings owing to line of contact.

- Cylindrical roller bearings are employed when maximum load carrying capacity is required in a given space. These consist of relatively short rollers that are positioned and guided by the cage. Advantages and disadvantages (refer Section 5.3)

- Spherical roller bearings are self-aligning rolling contact bearings in which balls are replaced by two rows of spherical rollers, which run on a common spherical surface in the outer race. Spherical roller bearings can carry relatively high radial and thrust loads. These permit minor angular misalignment of shaft relative to the housing.

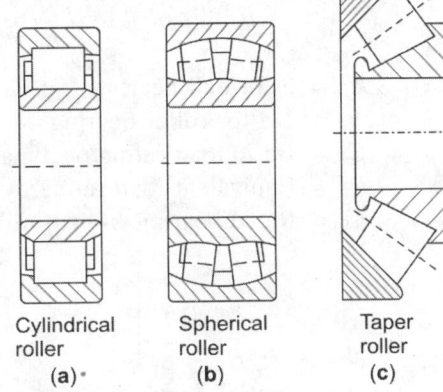

| Cylindrical roller | Spherical roller | Taper roller |
| (a) | (b) | (c) |

Fig. 5.3 Roller bearings (cylindrical, spherical and taper)

- Taper roller bearing consists of rolling elements in form of frustum of a cone, arranged in a way such that axes of rolling elements intersect in a common apex point on bearing axis, thereby promoting pure rolling motion between conical surfaces (for advantages and disadvantages refer Section 5.3).

5.6 SELECTION OF RADIAL BALL BEARING AND ROLLER BEARING

Selection depends upon requirement of the situation and characteristics of different types of bearings and criteria is as follows:

1. If a misalignment between the axes of shaft and housing is likely, self-aligning ball bearing and spherical roller bearings are used.
2. Ball bearings are used for low and medium radial loads, whereas roller bearings are selected for heavy loads and large shaft diameters.
3. For medium thrust loads thrust ball bearings are used whereas for heavy thrust loads cylindrical roller thrust bearing is employed. For carrying thrust loads in either direction, double acting thrust bearings can be used.
4. For high speed applications, owing to dependence of speed on temperature rise of bearing; angular contact bearing, cylindrical roller bearing and deep groove ball bearings are suggested.
5. For supporting radial and thrust components of the load, spherical roller bearing; angular contact bearing and deep groove ball bearing can be employed.

6. For low noise or quiet operations, deep groove ball bearings are recommended.
7. Taper roller bearings, double row cylindrical roller bearings may be employed, when rigidity is an important criteria in certain applications like machine tool spindles.

For proper selection of radial ball bearing and roller bearing, bearing characteristics/properties must match with the requirements of application and bearing design characteristics and application needs knowledge forms the core basis of selection criteria.

5.7 BEARING LIFE (L_{10})

Bearing life is defined as number of revolutions of bearing before first visible evidence of fatigue is seen in material of one of the bearings. Life of identical set of ball or roller bearings is defined as number of revolutions which 90% of group of bearings will complete before first evidence of fatigue develops.

Expected life of bearing:

$$L_{10} = \left(\frac{C}{P}\right)^k \times 10^6 \text{ revolution}$$

k = 3 for ball bearing
= 1/3 for roller bearing
C = Basic load rating/or dynamic load capacity (N)
P = Equivalent load rating/dynamic load
L_{10} = Rated bearing life (in million revolutions)

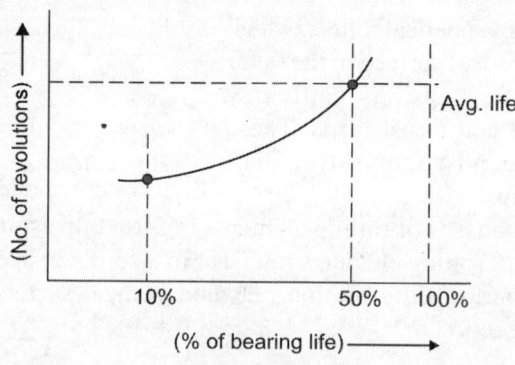

Fig. 5.4

50% group of bearing has 5 times the rated life.
So for all types of ball bearing

$$C = P(L_{10})^{1/3}$$

and for all types of roller bearings

$$C = P(L_{10})^{0.3}$$

Relationship between life in million revolutions and life in working hours is given by:

$$L_{10H} = \frac{L_{10} \times 10^6}{60n}$$

where L_{10H} = Rated bearing life (in hours)
n = Speed of rotation (rpm)

• Bearing life of an individual ball bearing can be defined as the number of revolutions (or hours of service at some given constant speed), that the bearing runs before first evidence of fatigue cracks in balls or races.

- Rating life of a group of apparently identical ball bearings is defined as a number of revolutions that 90% of the bearings will complete or exceed before the first evidence of fatigue crack.

Statistically it can be proved that the life, which 50% of a group of bearings will complete or exceed, is approximately five times the rating or L_{10} life. So for majority of bearings, actual life is considerably more than rated life.

5.8 DYNAMIC EQUIVALENT LOAD FOR ROLLER CONTACT BEARING UNDER CONSTANT AND VARIABLE LOADING

It is defined as the constant stationary radial or axial load, which if applied to a bearing with inner ring rotating would give the same life as that which the bearing will attain under actual conditions of load and rotation.

$$\boxed{F = XVF_r + YF_a}$$

where
- F = Equivalent dynamic load
- V = Velocity factor = 1 (when inner race rotates)
- F_r = Radial load = 1.2 (when outer race rotates)
- F_a = Axial load
- X = Radial factor and Y = Thrust factor

5.8.1 Static Bearing Capacity (C_0)

It is defined as the static radial load for radial ball bearing (or axial load for thrust ball bearing) which an application would produce deformation in balls and races of 0.1% of ball diameter and is denoted by C_0.

$$\boxed{C_0 = \left(\frac{kd^2 z}{s} \right)} \quad \text{(Stribeck's equation)}$$

where
- z = Number of balls
- d = Diameter of ball
- k = Factor dependent on modulus of elasticity of materials and radii of curvature at the point of contact.

5.8.2 Static Equivalent load (F_0)

It is defined as the static radial or axial load, which if applied would cause total deformation of ball and races contact which occurs under actual condition of loading.

$$F_0 = X_0 F_r + Y_0 F_a \text{ and } F_0 > F_r$$

where
- X_0 = Radial factor and Y_0 = Thrust factor
- F_r = Radial load and F_a = Axial load

5.8.3 Dynamic Bearing Capacity

It is defined as the dynamic load, which if applied constantly on bearing as radial load (or axial load), with stationary outer ring can give life of 1 million revolutions with 10% failure.

If loads $(F_1, F_2, F_3, ...)$ act on bearing with $(n_1, n_2, n_3, ...)$ revolutions, then the equivalent load

$$F_{eq} = \left[\frac{n_1 F_1^k + n_2 F_2^k + n_3 F_3^k + ...}{n_1 + n_2 + n_3} \right]^{1/k}$$

where
- k = 3 for ball bearing
- = 10/3 for roller bearing

5.8.4 Design for Variable Loading

In certain applications loads are not constant on ball bearings, but they are subjected to cyclic loads and speeds.

For example, radial load 'x_1'N for 'y_1' rpm for 'z_1'% of time.
and radial load 'x_2'N for 'y_2' rpm for 'z_2'% of time and so on.

Then we need to consider whole cycle for calculation of dynamic load capacity of the bearing.

For example, let $P_1, P_2, P_3, \ldots$ loads are acting for $n_1, n_2, n_3, \ldots$ speeds, then for 1st element

life 'L_1' is given as: $L_1 = \left(\dfrac{C}{P_1}\right)^3 \times 10^6$ rev.

So life consumed in 1 revolution

$$= \frac{1}{(C/P_1)^3 \times 10^6}$$

$$= \left(\frac{P_1}{C}\right)^3 \times \frac{1}{10^6}$$

So life consumed in N_1 revolutions

$$= \frac{N_1 P_1^3}{10^6 C}$$

Similarly life consumed in N_2 revolutions

$$= \frac{N_2 P_2^3}{10^6 C} \text{ and so on.}$$

So total life $= \dfrac{N_1 P_1^3}{10^6 C} + \dfrac{N_2 P_2^3}{10^6 C} + \ldots = \dfrac{N P_e^3}{10^6 C}$

(where P_e = Equivalent load for complete work cycle)

$$= \frac{N_1 P_1^3}{10^6 C} + \frac{N_2 P_2^3}{10^6 C} + \ldots = \frac{N P_e^3}{10^6 C}$$

$\Rightarrow \qquad N P_e^3 = N_1 P_1^3 + N_2 P_2^3 + \ldots$ (where $N = N_1 + N_2 + \ldots$)

$\Rightarrow \qquad P_e = \sqrt[3]{\dfrac{(N_1 P_1^3 + N_2 P_2^3 + \ldots)}{N_1 + N_2 + \ldots}}$

$$\boxed{P_e = \sqrt[3]{\frac{\sum N P^3}{\sum N}}}$$

If the load does not vary in steps of constant magnitude, but varies continuously with time then

$$P_e = \sqrt[3]{\frac{\int\limits_0^N P^3 \, dN}{\int\limits_0^N dN}} \qquad \Rightarrow \qquad P_e = \left[\frac{1}{N} \int\limits_0^N P^3 \, dN\right]^{1/3}$$

5.8.5 Dynamic Load Carrying Capacity (*C*)

It is defined as the radial load in radial bearings (or thrust load in thrust bearings) that can be carried for a minimum life of one million revolutions.

Minimum life is 'L_{10}' life that 90% of bearings will reach or exceed before fatigue failure.

Dynamic load carrying capacity is based on the assumption that the inner race is rotating while the outer race in stationary.

5.9 RELIABILITY OF BEARING

Reliability of bearing is generally defined as the ratio of number of bearings which have successfully completed '*L*' million revolutions to the total number of bearings under test.

It can also be defined as the chance that a rolling contact bearing will survive for a specified number of revolutions and is given by the relation.

$$R = \exp\left[-\left(\frac{x - x_0}{\theta - x_0}\right)^b\right]$$

where R = Reliability

x = Dimensionless variate = L/L_{10}

(where L_{10} = Life which at least 90% of group will achieve before failure criterion or rating life)

x_0 = Guaranteed or minimum value of variate

θ = Characteristic parameter corresponding to 63.2121 percentile value of variate x

By 1st definition of reliability (above)

i.e. $R = \dfrac{\text{No. of bearings which have successfully completed } L \text{ million revolutions}}{\text{Total no. of bearings under test}}$

= 90% or 0.9 as generally rating life of bearings is considered; life in million of revolutions which 90% of group of bearings achieve before failure occurs (which refers to criterion of around 6.55 mm² of pitting occur).

But for critical bearings or when there is risk of human life, reliability of more than 90% is required.

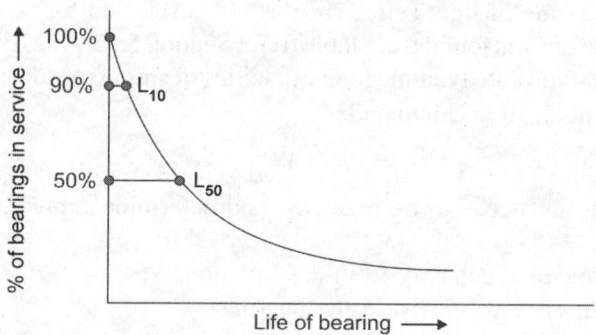

Fig. 5.5 Weibull distribution curve

For reliability to be other than 90%:

From graph, $L_{50} = 5L_{10}$

So $R = e^{-(L_H/a)^b}$ [L_H = Life of bearing in hours]

[*a* and *b* being constants]

(a = 6.84 and b = 1.7, applying $L_{50} = 5L_{10}$)

So
$$\frac{1}{R} = e^{(L_H/a)^b}$$

$$\Rightarrow \qquad \log_e\left(\frac{1}{R}\right) = \frac{(L_H)^b}{a}$$

or
$$\log_e\left(\frac{1}{R_{90}}\right) = \left(\frac{L_{10H}}{a}\right)^b \qquad\qquad [R_{90} = 0.9, \text{ also from } L_{50} = 5L_{10}]$$

So
$$\left(\frac{L_H}{L_{10H}}\right)^b = \left[\frac{\log_e\left(\frac{1}{R}\right)}{\log_e\left(\frac{1}{R_{90}}\right)}\right] \Rightarrow \boxed{\left(\frac{L_H}{L_{10H}}\right) = \left[\frac{\log_e\left(\frac{1}{R}\right)}{\log_e\left(\frac{1}{R_{90}}\right)}\right]^{1/b}}$$

Where L_{10H} is the life of bearings (90% of the lot remaining safe) in hours.

If there are 'N' bearings in the system and each bearing has an individual reliability of N, then reliability that 1 out of 'N' bearings fail during life time:

$$\boxed{R_S = (R)^N}$$

5.10 SELECTION OF ROLLING CONTACT BEARING

Selection procedure for rolling contact bearing from manufacture's catalogue is:

1. Radial and axial forces acting on the bearing and the diameter of shaft need to be determined for bearing to be fitted on the shaft.
2. Type of bearing for specific application needs to be selected from selection criteria.
3. Values of 'X' (radial factor) and 'Y' (thrust factor) are to be determined for specific application (from manufacturer's catalogue), for 'single row deep groove ball bearing'.
   'X and Y' depends on two ratios:
   (F_a/F_r) and (F_a/C_0), where 'C_0' is the static load capacity and F_a and F_r being the axial and radial loads respectively.
   For example, a bearing of light series such as 60, is selected for the given diameter of the shaft and the value of C_0 is found from table (refer Section 5.13, ASME and Indian Standards). [Dimensions and static and dynamic load capacities of single row deep groove ball bearings]
4. Equivalent dynamic load is calculated.

$$\boxed{P = XF_r + YF_a}$$

5. Expected bearing life needs to be practised and determine/express the life L_{10} in million revolutions.
6. Determine dynamic load capacity from the equations
   (for ball bearing)   (for roller bearing)

$$\boxed{C = P\left(L_{10}\right)^{1/3}} \qquad\qquad \boxed{C = P\left(L_{10}\right)^{0.3}}$$

where L_{10} = Rating life (million of revolutions)
$\qquad C$ = Dynamic load capacity (N)
$\qquad P$ = Equivalent dynamic load (N)

7. If the selected bearing of series (selected) has required dynamic capacity, selection criteria is fulfilled.

If not, then bearing of next consecutive series is selected and process is repeated from step 3 onwards.

Thus bearings (ball and roller) are selected by trial and error method as given above and is applicable to other type of bearings also.

5.10.1 Bearing Designation (Rolling Contact Bearing)

Rolling contact bearings are designated by 4 digits which are explained as follows:
1. Last two digits indicate bore diameter (in mm) divided by 5.
2. Second digit indicates bearing series which are specified as:
   (a) Extra light   (c) Medium
   (b) Light   (d) Heavy
3. First digit specifies type of rolling contact bearing, e.g. 6—deep groove ball bearing, etc.

For example, bearing designation '6,408' can be explained as—a 'deep groove ball' bearing of 'heavy series' with bore diameter of '8 × 5 = 40 mm'.

5.11 ASME AND INDIAN STANDARDS

Table 5.1 X and Y factors for single row deep groove ball bearing

$\left(\dfrac{F_a}{C_0}\right)$	$\left(\dfrac{F_a}{F_r}\right) \leq e$		$\left(\dfrac{F_a}{F_r}\right) > e$		e
	X	Y	X	Y	
0.025	1	0	0.56	2.0	0.22
0.040	1	0	0.56	1.8	0.24
0.070	1	0	0.56	1.6	0.27
0.130	1	0	0.56	1.4	0.31
0.250	1	0	0.56	1.2	0.37
0.500	1	0	0.56	1.0	0.44

Table 5.2 Dimensions and static and dynamic load capacities of single row deep groove ball bearings[4]

Principal dimensions (mm)			Basic load ratings (N)		Designation
d	D	B	C	C_0	
10	19	5	1,480	630	61,800
	26	8	4,620	1,960	6,000
	30	9	5,070	2,240	6,200
	35	11	8,060	3,750	6,300
12	21	5	1,430	695	61,801
	28	8	5,070	2,240	6,001
	32	10	6,890	3,100	6,201
	37	12	9,750	4,650	6,301
15	24	5	1,560	815	61,802
	32	9	5,590	2,500	6,002
	35	11	7,800	3,550	6,202
	42	13	11,400	5,400	6,302
17	26	5	1,680	930	61,803
	35	10	6,050	2,800	6,003
	40	12	9,560	4,500	6,202

(Contd...)

(Contd...)

Principal dimensions (mm)			Basic load ratings (N)		Designation
d	*D*	*B*	*C*	C_0	
	47	14	13,500	6,550	6,303
	62	17	22,900	11,800	6,403
20	32	7	2,700	1,500	61,804
	42	8	7,020	3,400	16,404
	42	12	9,360	4,500	6,004
	47	14	12,700	6,200	6,204
	52	15	15,900	7,800	6,304
	72	19	30,700	16,600	6,404
25	37	7	3,120	1,960	61,805
	47	8	7,610	4,000	16,005
	47	12	11,200	5,600	6,005
	52	15	14,000	6,950	6,205
	62	17	22,500	11,400	6,305
	80	21	35,800	19,600	6,405
30	42	7	3,120	2,080	61,806
	55	9	11,200	5,850	16,006
	55	13	13,300	6,800	6,006
	62	16	19,500	10,000	6,206
	72	19	28,100	14,600	6,306
	90	23	43,600	24,000	6,406
35	47	7	4,030	3,000	61,807
	62	9	12,400	6,950	16,007
	62	14	15,900	8,500	6,007
	72	17	25,500	13,700	6,207
	80	21	33,200	18,000	6,307
	100	25	55,300	31,000	6,407
40	52	7	4,160	3,350	61,808
	68	9	13,300	7,800	16,008
	68	15	16,800	9,300	6,008
	80	18	30,700	16,600	6,208
	90	23	41,000	22,400	6,308
	110	27	63,700	36,500	6,408
45	58	7	6,050	3,800	61,809
	75	10	15,600	9,300	16,009
	75	16	21,200	12,200	6,009
	85	19	33,200	18,600	6,209
	100	25	52,700	30,000	6,309
50	120	29	76,100	45,500	6,409
	65	7	6,240	4,250	61,810
	80	10	16,300	10,000	16,010
	80	16	21,600	13,200	6,010
	90	20	35,100	19,600	6,210
	110	27	61,800	36,000	6,310
	130	31	87,100	52,000	6,410
55	72	9	8,320	5,600	61,811
	90	11	19,500	12,200	16,011
	90	18	28,100	17,000	6,011

(Contd...)

(Contd...)

Principal dimensions (mm)			Basic load ratings (N)		Designation
d	D	B	C	C_0	
	40	12	9,560	4,500	6,202
	100	21	43,600	25,000	6,211
	120	29	71,500	41,500	6,311
	140	33	99,500	63,000	6,411
60	78	10	8,710	6,100	61,812
	95	11	19,900	13,200	16,012
	95	18	29,600	18,300	6,012
	110	22	47,500	28,000	6,212
	130	31	81,900	48,000	6,312
	150	35	108,000	69,500	6,412
65	85	10	11,700	8,300	61,813
	100	11	21,200	14,600	16,013
	100	18	30,700	19,600	6,013
	120	23	55,900	34,000	6,213
	140	33	92,300	56,000	6,313
	160	37	119,000	78,000	6,413
70	90	10	12,100	9,150	61,814
	110	13	28,100	19,000	16,014
	110	20	37,700	24,500	6,014
	125	24	61,800	37,500	6,214
	150	35	104,000	63,000	6,314
	180	42	143,000	104,000	6,414
75	95	10	12,500	9,800	61,815
	115	13	28,600	20,000	10,615
	115	20	39,700	26,000	6,015
	130	25	66,300	40,500	6,215
	160	37	112,000	72,000	6,315
	190	45	153,000	114,000	6,415

[4] In Table the following notations are used:
  d = Inner diameter of the bearing
  D = Outer diameter of the bearing
  B = Axial width of the bearing

WORKED EXAMPLES

1. A taper roller bearing has a dynamic load capacity of 26 kN. The desired life for 90% of the bearings is 8,000 hours and the speed is 300 rpm. Calculate the equivalent radial load that bearing can carry.

SOLUTION: As $L_{10} = \dfrac{60NL_{10H}}{10^6} = \dfrac{60(300)(8,000)}{10^6}$ [Bearing Life, Section 5.7]

$$= 144 \text{ million revolutions}$$

also $\qquad C = P(L_{10})^{0.3}$

$\Rightarrow$ $$P = \frac{C}{(L_{10})^{0.3}} = \frac{26,000}{(144)^{0.3}} = 5,854.16 \text{ N}$$

2. In a particular application the radial load acting on a ball bearing is 5 kN and the expected life for 90% of the bearings is 8,000 hours. Calculate the dynamic load carrying capacity of the bearing when the shaft rotates at 1,450 rpm.

SOLUTION: As $L_{10} = \dfrac{60NL_{10H}}{10^6}$ [Bearing life, Section 5.7]

$$= \frac{60 \times 1,450 \times 8,000}{10^6}$$

= 696 million revolutions

also $C = P(L_{10})^{1/3} = 5,000 \times (696)^{1/3} = 44,310.48$ N [Dynamic Load Capacity, Section 5.8.5]

3. A single row deep groove ball bearing is subjected to a pure radial force of 3 kN from a shaft that rotates at 600 rpm.

The expected life L_{10H} of the bearing is 30,000 hours.

The minimum acceptable diameter of the shaft is 40 mm. Select a suitable ball bearing for this application.

SOLUTION: Bearing is subjected to pure radial load

So $\qquad P = F_r = 3,000$ N

as $\qquad L_{10} = \dfrac{60NL_{10H}}{10^6} = \dfrac{60 \times 600 \times 30,000}{10^6} = 1,080$ million rev

also $\qquad C = P(L_{10})^{1/3} = (3,000)(1,080)^{1/3} = 30,779.57$ N

From tables (Ref: ASME and Indian Standards),

bearing no. '6,208' with dynamic load capacity ($C = 30,700$ N) is suitable for this application.

4. Select a deep groove ball bearing for a shaft of diameter 20 mm running at a speed of 1,000 rpm. The required bearing life is 5,000 hours at a reliability of 99%. The radial load is 2.5 kN and the axial load is 1.2 kN. The minimum operating temperature is 40 °C and inner race rotates.

SOLUTION: $\qquad$ Given: Inner diameter of bearing = 20 mm

$\qquad$ Speed (N) = 1,000 rpm

$\qquad$ Bearing life (L_H) = 5,000

$\qquad$ Hours at reliability of 99%

$\qquad$ Radial load (W_R) = 2.5 kN

$\qquad$ Axial load (W_A) = 1.2 kN

$\qquad$ Maximum temperature = 40 °C

$\qquad$ Inner race is rotating

Life of the bearing corresponding to 99% reliability

$$L_{99} = 60NL_{10} = 60 \times 1,000 \times 5,000 = 300 \times 10^6 \text{ revolutions}$$

Basic dynamic equivalent radial load

$$W = XVW_R \times YW_A$$

$$\frac{W_A}{W_R} = \frac{1.2}{2.5} = 0.48$$

which is greater than 0.44.

Here value of basic static load capacity (C_0) is not known, therefore, let us take $W_A/W_R = 0.5$. Now from table ('Sadhu Singh', Design Data Book), we find that values of 'X' and 'Y' corresponding to $\dfrac{W_A}{C_0} = 0.5$ and $\dfrac{W_A}{W_R} = 0.48$ (which is again greater than $e = 0.44$) are $X = 0.56$ and $Y = 1$. Since the rotational factor (V) for most of the bearings is 1, therefore, basic dynamic equivalent radial load

$$W = X V W_R + Y W_A$$
$$= 0.56 \times 1 \times 2.5 + 1 \times 1.2 = 2.6 \text{ kN} = 2{,}600 \text{ N}$$

For uniform and steady load, the service factor 'K_s' for ball bearing is 1.

Therefore, the bearing should be selected for $W = 2{,}600$ N

We know that the basic dynamic load rating

$$C = W \left(\frac{L}{10^6} \right)^{1/k}$$

(where $k = 3$ for ball bearing)

or $\qquad C = 2{,}600 \left(\dfrac{300 \times 10^6}{10^6} \right)^{1/3}$

or $\qquad C = 17{,}405 \text{ N} = 17.4 \text{ kN}$

From Design Data Handbook (Sadhu Singh), corresponding to

$C = 17.4$ kN and diameter = 2 mm.

On this basis, single row deep groove bearing number '6,404' should be the best option.

PREVIOUS YEAR UNIVERSITY QUESTIONS

1. Select a suitable roller bearing to carry a radial load of 10,000 N. The shaft rotates at 1,000 rpm, average life is 5,000 hours. Inner races rotates. Takes mild shock.   [UPTU 2005]

SOLUTION: Given:

Radial load $(W_R) = 10{,}000$ N
Speed $(N) = 1{,}000$ rpm
Average life $(L_H) = 5{,}000$ hours
Type of shock = Mild shock

Inner race rotates
Basic dynamic equivalent radial load

$$W = [XVW_R + YW_A]k_a$$

For mild shock $(k_a) = 1.5$
Equivalent load$(W) = [1 \times 1 \times 1{,}000] \times 1.5$
where $\qquad\qquad V = 1$ for most of bearings
and here $\qquad\quad X = 1$
$\qquad\therefore\ W = 15{,}000 \text{ N} = 1.5 \text{ kN}$
Life of bearing $(L) = 60 \times N \times L_H = 60 \times 1{,}000 \times 50{,}000$
$\qquad\qquad\qquad = 300 \times 10^6$ revolutions

We know that basic dynamic load rating

$$C = W \left(\frac{L}{10^6} \right)^{1/k} \text{ and } k = 10/3 \text{ (for roller bearing)}$$

$\Rightarrow \qquad C = 15 \left(\dfrac{300 \times 10^6}{10^6} \right)^{3/10} \text{ or } C = 83 \text{ kN}$

From Data Book, for $C = 83$ kN,
Roller bearing number 2,310 should be used.

2. Select a single row deep groove ball bearing for radial load of 4 kN and an axial load of 5 kN, operating at a speed of 1,500 rpm for an average life of 5 years at 10 hours/day. Assume uniform and steady load.   [UPTU 2008, 2009]

Solution: Given: $W_R = 4$ kN $= 4,000$ N
$$W_A = 5 \text{ kN} = 5,000 \text{ N}, N = 1,500 \text{ rpm}$$

Since the average life of the bearing is 5 years at 10 hours per day, therefore, life of the bearing in hours $L_H = 5 \times 300 \times 10 = 15,000$ hours (assuming 300 working days per year).

Life of the bearing in revolution
$$L = 60 \, NL_H = 60 \times 1,500 \times 15,000$$
$$= 1,350 \times 10^6 \text{ revolutions}$$

Dynamic equivalent radial load
$$W = XVW_R + YW_A$$
$$= 0.56 \times 1 \times 4,000 + 1 \times 5,000$$
$$= 7,240 \text{ N}$$

(x and y are find out from Data Book)

Basic dynamic load

$$C = W\left(\frac{L}{10^6}\right)^{1/k} = 7,240\left(\frac{1,350 \times 10^6}{10^6}\right)^{1/3}$$
$$= 80.017 \text{ kN}$$

From, Data Book, select bearing no. 315 which has $C_0 = 72$ kN
$$C = 90 \text{ kN}$$

$$W_A/C_0 = \frac{5,000}{72,000} = 0.07$$

$$X = 0.56 \text{ and } Y = 1.6 \text{ (from Design Data Handbook)}$$

Substituting these values
$$W = XVW_R + YW_A$$
$$= 0.5 \times 1 \times 4,000 + 1.6 \times 5,000$$
$$= 10,240 \text{ N}$$

$\therefore$ Basic dynamic loading

$$C = 10,240\left(\frac{1,440 \times 10^6}{10^6}\right)^{1/3} = 115,635 \text{ N}$$
$$= 115.635 \text{ kN}$$

So from Design Data Handbook, bearing no. 3,190 is selected, which have $C = 120$ kN.

3. Explain basic static load rating and basic dynamic load rating. Select a single new deep groove ball bearing for a radial load of 45 kN and axial load of 6 kN, operating speed of 1500 rpm for an average life of 5 years at 10 hours/day under uniform and steady load conditions.
                                                                                          [UPTU 2010]

Solution: Basic static load rating and basic dynamic load rating (refer Section 5.8)

Given: $F_r = 4.5$ kN, $F_a = 6$ kW
$$N = 1,500 \text{ rpm}$$

Average life of 5 years at 10 hours/day

$$L_{50H} = 5 \times 365 \times 10 = 18{,}250 \text{ hours}$$
$$R = 0.5 \text{ (average life)}$$

as $\quad L_{50H} = 4.481\, L_{10H}[\log_e (1/R)]^{1/1.5} \Rightarrow 18{,}250$

$$= 4.48\, L_{10H}\, [\log_e (1/0.5)]^{1/1.5}$$

$$L_{10H} = 5{,}201.17 \text{ hours}$$

Also $\quad C = W\left(\dfrac{L}{10^6}\right)^{1/k}$

$$k = 3 \text{ for ball bearing}$$

and $\quad L = \dfrac{60\,N L_{10H}}{10^6}$

$$= \dfrac{60 \times 1{,}500 \times 5{,}201.17}{10^6}$$

$\Rightarrow \qquad L = 468.1053 \text{ million revolutions}$

$\Rightarrow \qquad C = W\left(\dfrac{L}{10^6}\right)^{1/3} \Rightarrow 468.10 = \left(\dfrac{C}{14.56}\right)^3$

or $\qquad C = 112.81 \text{ kW.}$

So selected bearing is 6,315/6,316.

4. What is the rated life of a rolling contact bearing? Find the rated life of a 60 mm bore, light series ball bearing under a 6,000 N radial load at 600 rpm. The bearing rotates with the inner rings. There is no shock loading. [UPTU 2010]

SOLUTION: Rating life of a group of apparently identical bearings is defined as a number of revolutions (on the number of hours at a given constant speed) that 90% of a group of bearings will complete or exceed before the first evidence of fatigue failure develops. It is also known as L_{10} life or minimum life and is denoted L_{10}.

Given: $\qquad d = 60$ mm, $\quad F_r = 6{,}000$ N, $N = 600$ rpm

as $\qquad P_e = (V F_r) \qquad$ [where V is the velocity factor = 1 (here)]

$\Rightarrow \qquad P_e = 1 \times 6{,}000 = 6{,}000$

Corresponding to 60 mm bore, '6,012' bearing no. is selected and for this $C = 29.60$ kN.

$$L_{10} = \left(\frac{C}{P_e}\right)^3 = \left(\frac{29.60 \times 10^3}{6{,}000}\right)^3 = 120.06 \text{ million of revolutions}$$

5. A deep groove ball bearing has dynamic capacity of 20,000 N and is to operate on the following work cycle; radial load of 6,000 N at 200 rpm for 25% time, radial load of 9,000 N at 500 rpm for 20% of the time, radial load of 9,000 N at 500 rpm for 20% of the time and radial load of 3,500 N at 400 rpm for the remaining period. Assuming the loads are steady and the inner race rotates, find the average expected life of the bearing in hours.

[UPTU 2011]

SOLUTION: Considering the work cycle of one minute duration, the values of load P and revolution N are tabulated as follows:

Element no.	$\dfrac{P}{(N)}$	Element time (minute)	Speed (rpm)	Revolutions N (in element time)
1	6,000	0.25	200	50
2	9,000	0.20	500	100
3	3,500	0.55	400	220
Total		**1.00**		**370**

Now for variable loading

$$P_e = \sqrt[3]{\dfrac{N_1 P_1^3 + N_2 P_2^3 + N_3 P_3^3}{N_1 + N_2 + N_3}}$$

$$= \sqrt[3]{\dfrac{50 \times (6,000)^3 + 1,000 \times (9,000)^3 + 220 \times (3,500)^3}{(50 + 100 + 220)}} = 6,313.93 \text{ N}$$

According to the load life relationship

$$L = \left(\dfrac{C}{P_e}\right)^3 = \left(\dfrac{20,000}{6,313.93}\right)^3 = 31.78 \text{ million rev}$$

$$L_n = \dfrac{L \times 10^6}{80\,N} = \dfrac{31.78 \times 10^6}{60 \times (370)} = 1,431.65 \text{ hours}$$

6. Select a suitable bearing for a 40 mm shaft that has to operate 8 hours/day, 5 days/week for 5 years and is to carry a stationary radial load of 2,500 N at 1,500 rpm. The use involves minor shock and inner ring is rotating. [UPTU 2011]

SOLUTION: Life of bearing in hours

$$L_H = 5 \times 260 \times 8 = 10,400 \text{ hours}$$

Life of the bearing in revolutions

$$\begin{aligned} L &= 60\,NL_H \\ &= 60 \times 1,500 \times 10,400 \\ &= 936 \times 10^6 \text{ revolutions} \end{aligned}$$

So $C = PL^{1/3} = 2,500 \times (936)^{1/3} = 24,454.86 \text{ N}$

'D' is the shaft diameter which is given by 40 mm.

So from Data Hand Book, $C = 6,207$.

5.12 LUBRICATION OF BALL AND ROLLER BEARING

Lubrication is required to reduce or prevent friction, wear, heat generated, corrosion and protection from dirt and other foreign particles. Lubrication is done by means of oils and greases.

Advantages of grease

a. Less maintenance
b. No leakage
c. Simple housing design
d. Better sealing against rust

Advantages of oil

a. Easy feed into contact areas
b. More effective in carrying frictional heat
c. More effective in flushing out dirt, corrosion and foreign particles outside the bearings

Criteria

- If operating temperature for bearing is less than (<) 100 °C, grease is preferred; and if it is greater than (>) 100 °C, oils are preferred.
- If bore size (mm) × Speed (rpm) < (less than) 200,000 → grease is preferred, and if it is > 200,000 → oil is preferred.
- For high speed, heavy duty applications: Oils are preferred and for low and moderate loads grease is preferred, as it provides cheaper and simple mode of lubrication. If there is a central lubricating system as in gear box the same lubricating oil is used for bearings too.

5.13 MOUNTING OF BEARING

Inner race is interference fits over the shaft and outer race is also interference fits over the bearing housing.

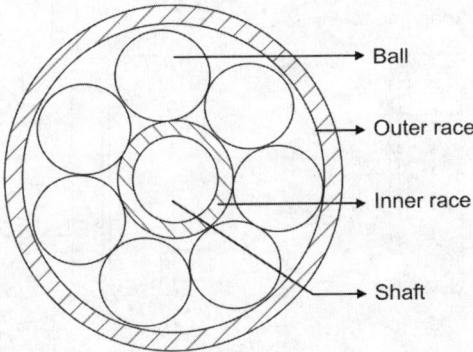

Fig. 5.6 Ball bearing

Inner race should be tightly fit to prevent the relative motion between inner race and shaft and at the same time, it should not be too tight, as to cause deformation of inner race and destroying clearance between rolling elements and races.

Outer race should also be tight, optimum over bearing housing (to a lesser degree than inner race), which if not may cause 'creep', thereby implying slow rotation of outer race relative to its seating.

When two bearings are mounted on the same bearing, then outer race of one of them should be permitted to shift axially to take care of axial deflection of shaft caused by thrust load or temperature variation.

Basic principle or requirements of mounting is to restrict the displacement of inner as well as outer races in axial direction by positive means.

For example, in figure, the bearing is press fitted on the adaptor sleeve (having a small taper).

Because of taper displacement of inner race to the left side is restricted, a washer and lock nut is provided to restrict the displacement of inner race to the right side.

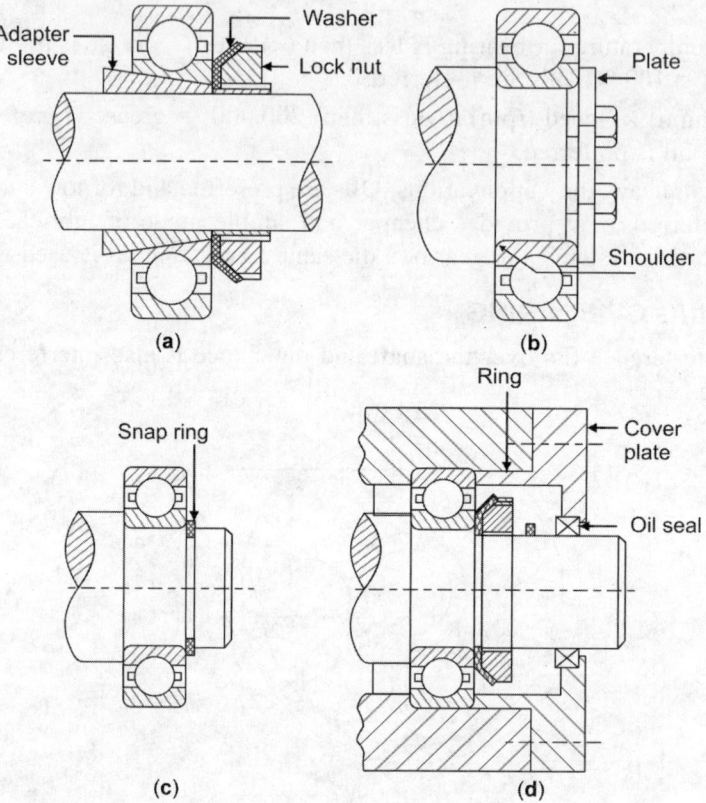

Fig. 5.7 Mountings of bearing

KEY TERMS (SUMMARY)

- Bearing classification:
  - Depending upon direction of force:
    - o Radial bearing
    - o Thrust bearing
  - Depending upon type of friction:
    - o Rolling contact
    - o Sliding contact
    - o Types of rolling contact bearings

For starting conditions and at moderate speeds the frictional loss in rolling contact bearing are lower than that of equivalent hydrodynamic journal bearings.

Types:
- o Deep groove ball bearing
- o Cylindrical roller bearing
- o Angular contact bearing
- o Self-aligning bearing
- o Taper roller bearing
- o Thrust ball bearing
- o Selection criteria of bearing

  Low/medium radial load   –   Ball bearing

Heavy load	–	Roller bearing
Misalignment	–	Self-aligning bearing
Medium thrust	–	Thrust ball bearing
Heavy thrust	–	Cylindrical thrust bearing
Load in either direction	–	Double acting thrust bearing
Radial and thrust load bath	–	Deep groove bearing, angular contact bearing, spherical roller bearing
High speed	–	Deep groove ball bearing, angular contact bearing
Rigidity	–	Double row cylindrical roller bearing, taper roller bearing
Less noise	–	Deep groove ball bearing

- Static load carrying capacity

  Load acting on bearing when shaft is stationary is static load and its value which corresponds to a total permanent deformation of ball and races at the most heavily stressed point of contact being equal to 0.0001 of ball diameter is "static load" carrying capacity.

- Dynamic load carrying capacity

  Radial load in radial bearing or axial load in thrust bearing that can be carried for a minimum life of one million revolutions, assuming inner is rotating and outer race is stationary.

- Equivalent bearing load

  Constant radial load in radial bearing or axial load in axial bearing, which if applied to bearing would give same life as that of which the bearing will attain in actual conditions.

  $$F_{eq} = XVF_r + YF_a$$

  where F_{eq} = Equivalent dynamic load

  F_r = Radial load

  F_a = Axial load

  X = Radial factor

  Y = Thrust factor

  V = Race – Rotation factor = 1 (when inner race rotates)

  = 1.2 (when outer race rotates)

- Bearing life under varying load

  $$L = L = \left(\frac{C}{F_e}\right)^k \text{ million revolutions}$$

  where C = Dynamic load capacity

  F_e = Equivalent dynamic load

  k = 3 (for ball bearing) and 10/3 (for roller bearing)

- Designation of ball bearing

  'PQRS'

  where P = Bearing type

  Q = Bearing series (1—extra light, 2—light, 3—medium and 4—heavy)

  RS = When multiplied by 5, it gives shaft diameter in mm.

- Bearing Failure:
  o Abrasive wear
  o Corrosive wear
  o Pitting
  o Scoring

REVIEW QUESTIONS

Short Answers

1. Discuss why rolling element bearings are also called antifriction bearings.
2. Give an account of classification of rolling element bearings indicating their applications.
3. Define the terms: Static bearing capacity; dynamic bearing capacity; equivalent load; cubic mean load; and life of bearing (in million of revolutions).
4. Compare ball and roller bearings.

Long Answers

1. Indicate with proper justification of the typical solutions for which the following types of antifriction bearing are to be used exclusively.
   a. Deep groove ball bearing
   b. Roller bearing
   c. Needle bearing
   d. Angular contact ball bearing
   e. Thrust ball bearing
2. Enumerate the advantages and disadvantages of rolling contact bearings and sliding contact bearings and selection criteria.
3. Explain how the following factors influence the life of a bearing: Load, speed and temperature.
4. What are the main components of rolling contact bearing?

Numericals

1. A shaft rotating at 1,440 rpm is supported by two bearings. The forces acting on each bearing and 6 kN radial load and 3.5 kN axial thrust.
   If the shaft diameter is 40 mm and the expected life of bearing is 500 hours, suggest a suitable bearing for the application.

   [bearing no. 6,308 (deep groove ball) or self-aligning (2,308 K)]
2. Select a suitable rolling element bearing with the following available data:
   Radial load = 2,500 N
   Axial load = 600 N
   Speed = 100 rpm
   Expected life = 2 years at 8 hours/day run
   Operating temperature = 100 °C
   Shaft diameter = 50 mm.
3. Select a deep groove ball bearing for a shaft of diameter 20 mm running at a speed of 1,000 rpm. The required bearing life is 5,000 hours at a reliability of 99%. The radial load is 2.5 kN and the axial load is 1.2 kN. The maximum operating temperature is 40 °C and inner race rotates.   [deep groove ball bearing 6,404]
4. Select a suitable roller bearing to carry a radial load of 10,000 N. The shaft rotates at 1,000 rpm, average life is 5,000 hours. Inner race rotates. Take mild shock.

   [Roller bearing number—2,310]

6

IC Engine Parts

6.1 INTRODUCTION

Internal combustion engine is one of the most widely used prime moves operating on fossil fuel. These engines can be classified as:

- According to number of strokes: 2 stroke and 4 stroke engines.
- On the basis of operating cycle: Petrol engines (based on otto cycle) and diesel engines (based on diesel or dual cycle)
- According to arrangement of cylinders: Low speed engines (with horizontal cylinders) and high speed engines (with vertical inline arrangement)
  For compact, shorter and lighter constructions: *V*-type and radial engines.
  An IC engine comprises following main components:

  - Cylinder block
  - Cylinder liner
  - Cylinder head
  - Piston
  - Piston rings
  - Piston pin

135

- Connecting rod
- Crankshaft

Other parts like camshaft, valve operating mechanisms, main bearings, etc.

6.2 SELECTION OF TYPE OF IC ENGINE

Refer Section 6.1 (Classification of IC Engine).

6.3 GENERAL DESIGN CONSIDERATIONS

Design considerations for different components of IC engine are as follows:

Cylinder

Cylinder of an IC engine acts as the structured member and retains the working fluid in a closed space with movable piston. Very small engines, consist of a single cast piece comprising cylinder, water jacket and frame whereas small engines have cylinder and jacket cast as one piece.

For wear and tear, avoid once, separate liners which are more economical and are easily replaceable.

Cylinder is tested to high explosive pressure which is approximately 3–8 times the maximum compression pressure and high temperature ranging between 1,800 K to 2,400 K. Thus the cylinder of an interval combustion engine should be able to withstand high working pressure and should be able to transfer the heat efficiently without thermal distortion taking place.

Cylinder liners owing to placed about cylinder walls to avoid wear and tears are subjected to extreme conditions and must possess following qualities:

1. Should have corrosion resistance.
2. Strong enough to resist gas pressure and thermal stresses.
3. Sufficiently hard to resist wear.
4. No distortion of inner surface, under extreme pressure and temperature conditions.
5. Symmetrical in shape to avoid unequal deflection and expansion due to gas load and thermal load respectively.

A cylinder liner which does not come in direct contact with the cooling medium is called the dry liner, whereas a wet liner comes in contact with the cooling medium. Materials considered for cylinder liner are—nickel cast iron, nickel chrome cast iron, cast steel, etc. Heat treatment is done for inner surface of liner to obtain a hard surface and chrome plating to obtain a smooth surface.

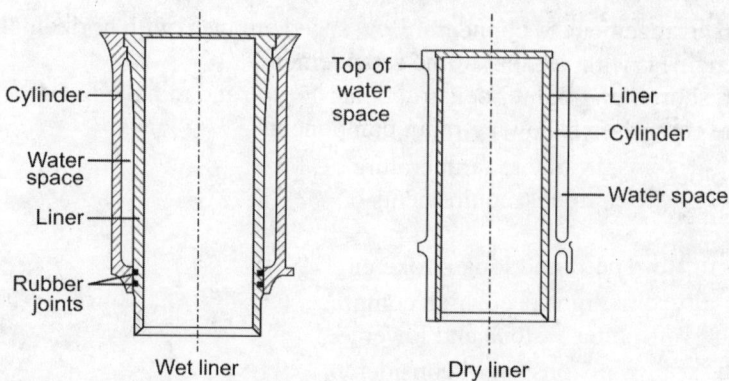

Fig. 6.1 Cylinder liners

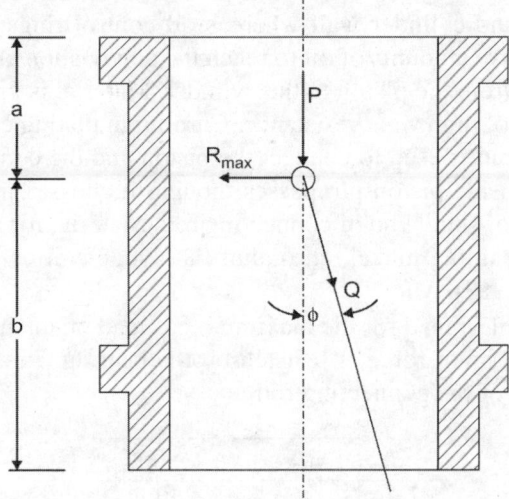

Fig. 6.2 Forces acting on cylinder liner

Piston

Piston being the most stressed element of all components of assembly, experiences high gas load, along with inertial and heat load. Piston must be strong enough to take these heat and pressure forces million of times during its lifetime.

Pistons are of two types:
- Trunk or full skirt type
- Short skirt type open at one end

Design criteria of piston requires it to slide with minimum friction, with minimum weight to keep down the inertial forces, and provide adequate heat path to prevent excessive heat temperature.

Design should be, so as to prevent the escape of burnt gases. Construction should be rigid enough to withstand thermal and mechanical distortions. A piston comprises 4 parts:
- Crown or head carrying gas force and major portion of heat load.
- Skirt to contain side thrust.
- Piston rings to create proper compression.
- Piston pin to connect to the connecting rod.

Head of piston needs to hot, whereas skirt is comparatively cool, otherwise thermal efficiency gets reduced.

Pistons are designed with tapered skirt to allow for the difference in expansion of upper and lower parts of the piston skirt due to temperature differential.

Depending upon the functional requirements of piston following materials are employed as piston materials:
- Cast iron: For low speed and long stroke engines like diesel engines.
- Aluminum alloy: For high speed, high compression short stroke comprising Si, Cu and Ni engines. Also for lighter pistons and lower (and Mg as alloying elements) inertial loads.

Forged aluminum alloy pistons are of considerable more strength and less porosity than cast iron piston.

Piston rings are of 2 types: Compression and oil control rings. Compression rings provide sealing between piston and cylinder wall whereas oil control rings maintain piston cooling effect and prevent excessive amounts of oil to reach the combustion chamber. Piston rings also help in transferring heat from the piston to the cylinder. Material is grey cast iron or alloy cast iron due to requirements of high wear resistance. Chromium plating can also be applied piston pin also called gudgeon pin is used to connect the piston and the connecting rod. It is usually hollow and tapered on inside piston pin passes though the bosses provided on the inside of piston skirt and the bush of small end of connecting rod. Material for piston pin is usually case hardened alloy steel containing nickel, chromium, molybdenum or vanadium having tensile strength from 710 MPa to 910 MPa.

Piston pin needs to be designed for the maximum gas load on the inertia force of the piston, whichever is larger. Maximum force is resisted by the bearing pressure between the piston, piston pin and small end of the connecting rod.

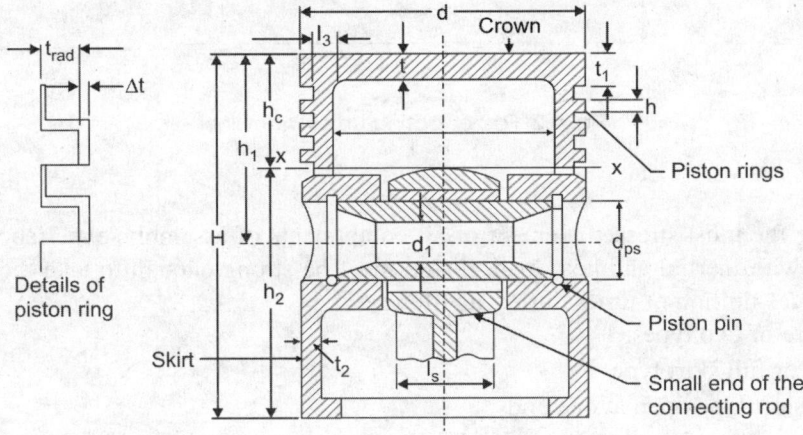

Fig. 6.3 Piston of IC Engine

Connecting Rod

Connecting rod is the intermediate member between the piston and crankshaft, with primary function being to transmit the push and pull from the piston pin to the crank pin and convert reciprocating motion of piston into rotary motion of crank.

Connecting rod structure comprises a long shank, a small end and a big end. Shank's cross-section may be I-section (most common), H-section, circular, etc.

Length of connecting rod is usually kept 3–4.5 times crank radius. Shorter length increases side thrust on cylinder whereas longer length increases engine height.

Material for connecting rods must have higher strength, and hence steels containing chromium, Cr-W, Cr-Ni and Mb with 0.4% carbon.

For carburetor engines connecting rods are made of steels containing 0.4–0.45% carbon and 1.4–1.8% Mn.

Connecting rods need improved fatigue strength owing to their fluctuating load conditions, and for this, connecting rods after press forging undergo machining and thermal treatments like normalizing, hardening, polishing and tempering.

For operation under extremely high loads, rods are made of titanium (instead of steel), which has excellent strength with light weight.

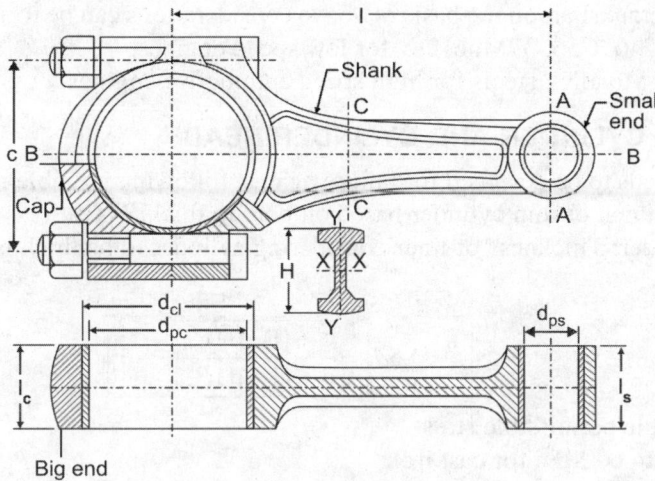

Fig. 6.4 Connecting rod

Stresses set up in connecting rod account from:

a. Gas pressure and piston inertia

b. Inertia of connecting rod

c. Friction of rings and piston

d. Frictional force in big and small end bearings

Crankshaft

Crankshaft is a shaft with a rotating member (crank) to transfer reciprocating motion into rotary one or vice versa. It is subjected to cyclic loads due to gas pressure, inertial forces and their couples.

Also, in addition to this bending moment is caused by weight of the flywheel or weight of rotor coupled to the engine thereby giving rise to torsional vibration of the shaft.

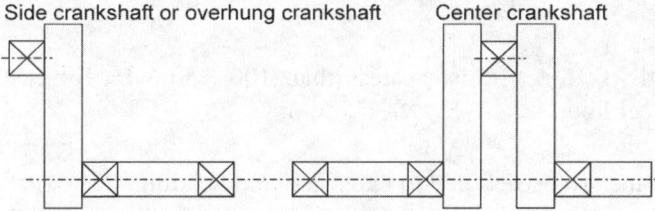

Fig. 6.5 Crankshafts

Crankshaft's popular design can be modeled as a simply supported beam with one or more spans between the supports (as shown in figure).

Crankshaft with only one center crank is called a single throw crankshaft and with more are called multithrow crankshafts.

A crankshaft needs to be strong enough to resist fluctuating loads and shock forces and rigid enough to keep the deflection and distortion under permissible limits.

Crankshaft can be classified into following two categories on the basis of position of crank.

• Side/overhung crankshaft

• Center crankshaft

Material of the crankshaft on the basis of above considerations can be forging from medium carbon steels like C40, C50, 37Mn6, etc. for low speed engines.

Cr-Ni, Cr-V, Cr-Mb alloy steels for high speed automotive engines.

6.4 DESIGN OF CYLINDER AND CYLINDER HEAD

A cylinder liner needs to be designed for failure against following possible modes for cylinder considering either thick or thin cylinder based on bore to thickness ratio.

a. Thick cylinder: Thickness of liner considering cylinder to be thick is computed by the equation

$$t = 0.5d \left[\sqrt{\frac{\sigma + (1-\mu)P_{max}}{\sigma - (1+\mu)P_{max}}} - 1 \right]$$

where σ is the permissible stress

and 50 to 60 MPa for cast iron

and 80 to 100 MPa for steel

d is the cylinder bore.

P_{max} is the maximum pressure.

μ is Poisson's ratio.

b. Thin cylinder: For thin cylinder, thickness of cylinder liner is computed by the relation

$$t = \frac{P_{max} \times d}{2\sigma_0} + C$$

where σ_0 is the permissible hoop stress and C is the reboring allowance (1.5 – 15.0 mm).

1. Taking on account of temperature stresses, a cylinder liner should be checked for thermal stress caused by high temperature difference between the outer and inner surfaces of the liner.

$$\sigma_{th} = \frac{E \propto \Delta T}{2(1-\mu)}$$

where $\Delta T = 100 - 150$ °C for top portion of liner

μ = Poisson's ratio

E = Modulus of elasticity

2. Total stress induced ($\sigma_0 + \sigma_{th}$) must be less than 100–130 MPa for cast iron and 180–200 MPa for steel liner.

3. Liner is checked for the total stress due to extension and bending, i.e. $\boxed{\sigma_L = \dfrac{P_{max} \times d}{4t}}$

4. Considering that liner is supported at two points and maximum side thrust R_{max} acts at a distance 'a' from the 'TDC (top dead center)' position of the piston, the bending moment

$$M = R_{max} \times \left(\frac{ab}{a+b} \right)$$

where a is the distance between the piston pin axis and the TDC position, b is the distance between the piston pin axis and the BDC position.

Bending stress (σ_b) = M/Z

where

$$Z = \frac{\pi}{32} \left[\frac{d_0^4 - d^4}{D_0} \right]$$

where d_0 and d are the outer and inner diameters of the cylinder lines with $d_0 = (d + 2t)$
The total stress due to extension and bending in the wall is $\sigma = (\sigma_L + \sigma_b)$ and this should
not exceed 60 MPa for cast iron and 110 MPa for steel liner.

Dimensions of cylinder block (cylinder head) can be determined empirically as follows:

a. Thickness of cylinder block wall:

$$t_1 = (0.045d + 2)\,\text{mm}$$

b. Thickness of cylinder flange $(t_2) = (1.2 - 1.4)t_1$

c. Thickness of jacket wall

$$t_3 = (0.032d + 1.5)\ \text{mm}$$

d. Water space between the outer cylinder wall and the inner jacket wall
   $t_4 = (0.08d + 6.5)$ mm.

   Cylinders are attached to the crank case by means of the flange, studs and nuts.

   Diameter of the studs may be obtained by the relation $(d_{\text{bolt}}) = d\left[\dfrac{P_{max}}{N\sigma_t}\right]^{0.5}$

   where N = Number of studs $= (0.25 - 0.5)\dfrac{d_p}{10} + 4$

       d_p = Pitch circle diameter (mm)

       σ_t = Allowable tensile strength of the bolt material

e. For cylinder head which becomes complicated due to the presence of inlet and
   exhaust valves, spank plug, fuel injector and the shape of the combination chamber.
   Thus for approximation, cylinder head may be assumed as a flat circular plate held
   rigidly at the circumference by a suitable number of studs. Thickness of this plate
   can be computed by:

$$t_h = d\left[\frac{C_{p(max)}}{\sigma}\right]^{0.5}$$

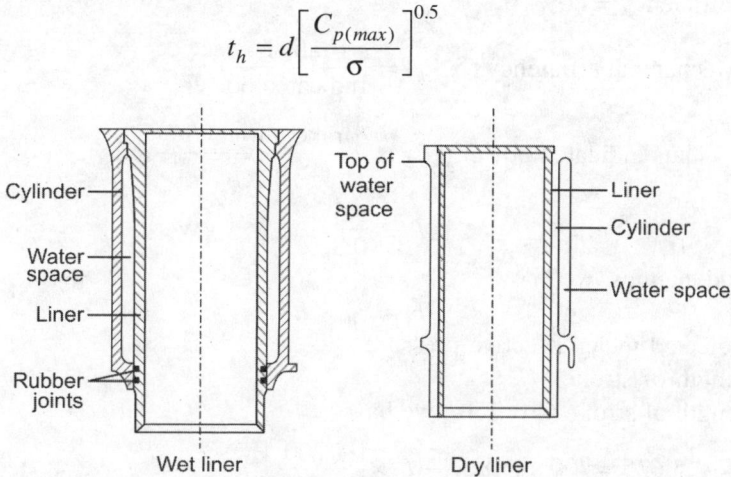

Fig. 6.6 Cylinder liner

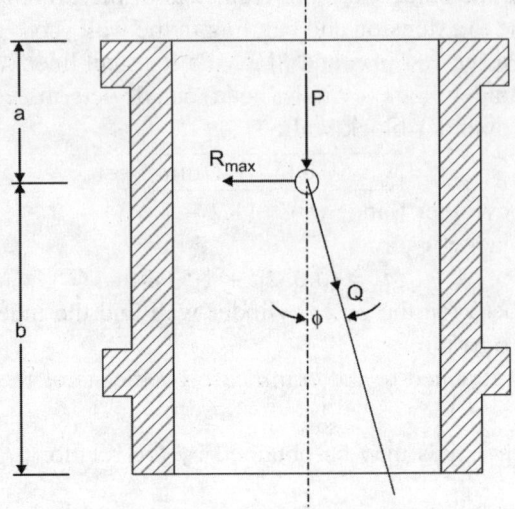

Fig. 6.7 Forces acting on cylinder

WORKED EXAMPLES

1. Determine the principal dimensions of cylinder for a vertical four stroke compression ignition engine from the following data:

Brake power = 4.5 kW, Speed = 1,200 rpm

Indicated mean effective pressure = 700 kN/m²

Mechanical efficiency = 80%

SOLUTION: Mechanical efficiency $(\delta_{mech}) = \dfrac{\text{Brake power}}{\text{Indicated power}}$

Thus indicated power $(IP) = \dfrac{\text{Brake power}\,(BP)}{\eta_{mech}}$

$$IP = \frac{4.5}{0.8} = 5.625 \text{ kW}$$

From thermodynamics, we know

$$IP = P_{in} \times LAN$$

where P_{in} = Indicated mean effective pressure

l = Length of stroke

Assuming length of stroke $(l) = 1.1d$, we have

$$5.625 = 700 \times 1.1d \times \frac{\pi}{4}d^2 \times \left[\frac{1,200}{2 \times 60}\right]$$

So $$d = \sqrt[3]{\frac{5.625 \times 2 \times 60 \times 4}{700 \times 1.1 \times \pi \times 1,200}} = 0.097 \text{ m or 97 mm}$$

$$l = 1.1 \times 97 = 106.7 \text{ mm} \approx 107 \text{ mm}$$

Assuming that the maximum explosion pressure

$$P_{max} = 8 \times P_{in} = 8 \times 700 = 5,600 \text{ kN/m}^2$$
$$= 5.6 \text{ N/mm}^2$$

For cast iron cylinder, allowable strength (σ) = 60 N/mm^2
and Poisson's ratio (μ) = 0.21
Thickness of the cylinder with reboring allowance is given by

$$t = 0.5d \left[\left[\frac{\sigma + (1-\mu)P_{max}}{\sigma - (1+\mu)P_{max}} \right]^{0.5} - 1 \right]$$

$$= 0.5 \times 97 \left[\left(\frac{60 + (1-0.21)5.6}{60 - (1+0.21)5.6} \right)^{0.5} - 1 \right]$$

$$= 4.85 \text{ mm say 6 mm.}$$

1. The hoop stress produced in the cylinder

$$\sigma_0 = \frac{P_{max}d}{2t} = \frac{5.6 \times 97}{2 \times 6} = 45.26 \text{ N/mm}^2$$

2. Thermal stresses $\left(\sigma_{th} = \frac{E \propto \Delta T}{2(1-\mu)} \right)$

   Let $\qquad \alpha = 11 \times 10^{-6}$ mm/°C and $\Delta T = 120$ °C

   $\therefore \qquad\qquad \sigma_{th} = \dfrac{1 \times 10^5 \times 11 \times 10^{-6} \times 120}{2(1-0.21)}$

   $$= 83.5 \text{ N/mm}^2$$

   Total stress ($\sigma_0 + \sigma_{th}$) = 45.26 + 83.5 = 128.76 N/mm^2.
   This value is acceptable as it is less than the tensile strength of cast iron (130 N/mm^2).

3. Longitudinal tensile stress

$$\sigma_L = \frac{P_{max}d}{4t} = \frac{5.6 \times 97}{4 \times 6} = 22.63 \text{ N/mm}^2$$

4. The side thrust in the cylinder
$$R = Q \sin \phi = P \tan \phi$$

   Assuming that for small obliquity of the connecting rod the maximum side thrust is 10% of the gas force,

We have

$$R_{max} = 0.1 \times \frac{\pi}{4} \times 97^2 \times 5.6 = 4,136 \text{ N}$$

Length of the cylinder= 1.25 × Stroke length
$$= 1.25 \times 107 = 133.75 \text{ mm}$$
Let the position of the piston pin from *TDC* be
$$a = 50 \text{ mm and } b = 70 \text{ mm}$$
Therefore, the bending moment is

$$M = R_{max} \times \frac{ab}{a+b}$$

$$= 4,136 \times \frac{50 \times 70}{50 + 70} = 120,633.3 \text{ N-mm}$$

Section modulus

$$Z = \frac{\pi}{32} \left[\frac{d_0^4 - d^4}{D_0} \right]$$

where $d_0 = d + 2t = 97 + 2 \times 6 = 109$ mm.

Therefore,
$$Z = \frac{\pi}{32}\left[\frac{10.9^4 - 97^4}{109}\right]$$
$$= 47,421.2 \text{ mm}^3$$

Bending stress $(\sigma_b) = \frac{M}{Z} = \frac{120,633.3}{47,421.2} = 2.5 \text{ N/mm}^2$

Total tensile stress $= \sigma_b + \sigma_L = 2.5 + 22.63$
$$= 25.13 \text{ N/mm}^2$$

which is less than allowable strength. Hence the design is satisfactory.

Other Dimensions

i. Thickness of the cylinder block wall
$$t_1 = 0.045d + 2 = 0.045 \times 97 + 2$$
$$= 6 \text{ mm}$$

ii. Thickness of the cylinder flange
$$t_2 = 1.3t_1 = 1.3 \times 6 = 8 \text{ mm}$$

iii. Water space $(t_4) = 0.08d + 6.5$
$$= 0.08 \times 97 + 6.5 = 14.26 \text{ mm} \approx 14 \text{ mm}$$

Assuming that the cylinder is fitted to the crank case with studs, the pitch circle diameter of studs

$$d_p = 1.6d, \text{ say } 140 \text{ mm}$$

Number of studs
$$(N) = (0.25 \text{ to } 0.5)\,\frac{d_p}{10} + 4$$
$$= 0.35 \times \frac{140}{10} + 4$$
$$= 8.9, \text{ say } 10$$

Cone diameter of the stud $(d_{bolt}) = d\left[\dfrac{P_{max}}{N\sigma_t}\right]^{0.5}$

Let us assume that the allowable strength of studs
$$\sigma_t = 80 \text{ N/mm}^2, \text{ then}$$
$$d_{bolt} = 97\left[\frac{5.6}{10 \times 80}\right]^{0.5} = 8.11 \text{ mm, say 10 mm nominal.}$$

Size:
Thickness of the cylinder head

$$t_h = d\left[\frac{P_{max}}{\sigma}\right]^{0.5}$$
$$= 97\left[\frac{0.162 \times 5.6}{60}\right]^{0.5}$$
$$= 11.92 \text{ mm, say } 12.5 \text{ mm}$$

2. Determine the thickness of a cast iron cylinder wall and the stresses for a 250 mm petrol engine with a maximum gas pressure of 3 N/mm². Take the reboring factor for the cylinder wall as 7.5 mm and Poisson's ratio as 0.25 for cylinder material. Take maximum hoop stress as 45 MPa for the material.

SOLUTION: Given:

Engine cylinder bore $(d) = 250$ mm

Maximum gas pressure $(P) = 3$ N/mm^2

Reboring factor $(C) = 7.5$ mm.

Poisson's ratio $(D) = 0.25$ for cylinder material

Permissible or maximum hoop stress $(\sigma_c) = 45$ MPa

Determine cylinder with thickness 't'.

Thickness of cylinder wall:

The thickness of a cylinder wall 't' may be found by using their cylindrical formula, i.e.

$$t = \frac{P \times d}{2\sigma_c} + C = \frac{3.0 \times 250}{2 \times 48} + 7.5 = 15.83 \text{ mm}$$

$$= 16 \text{ mm}$$

Stresses in Cylinder Wall

The cylinder wall is subjected to gas pressure and the piston side thrust. The gas pressure produces following two types of stress:

   i. Longitudinal stress ii. Circumferential or hoop stress

Longitudinal stress

$$T_l = \frac{\pi/4 \times d^2 \times P}{\pi/4 \left(d_0^2 - d^2\right)} = \frac{d^2 P}{d_0^2 - d^2}$$

$$d_0 = d + 2t = 250 + 2 \times 16 = 282 \text{ mm}.$$

$$\sigma_l = \frac{(250)^2 \times 3.0}{(282)^2 - (250)^2} = 11.0138 \text{ N/mm}^2 = 11.014 \text{ N/mm}^2$$

Circumferential stress or hoop stress

$$\sigma_c = \frac{dP}{2t} = \frac{250 \times 3.0}{2 \times 16} = 23.4375 \approx 23.44 \text{ N/mm}^2$$

Net longitudinal stress

$$\left(\sigma_l\right)_{net} = T_l - \frac{T_c}{m} = 11.014 - 23.44 \times 0.25 = 5.154 \text{ N/mm}^2$$

Net circumferential stress

$$\left(\sigma_c\right)_{net} = T_c - \frac{T_l}{m} = 23.44 - 11.014 \times 0.25$$

$$= 20.6865 = 20.69 \approx 21 \text{ N/mm}^2$$

PREVIOUS YEAR UNIVERSITY QUESTIONS

1. A 4 stroke diesel engine has the following specifications:

Brake power = 5 kW

Speed = 1,200 rpm

Indicated mean effective pressure = 0.35 N/mm^2

Mechanical efficiency = 80%

Determine the bore and length of the cylinder and thickness of cylinder head. (UPTU 2009)

SOLUTION: Given: $BP = 5$ kW (brake power)

$\qquad N =$ Speed $= 1,200$ rpm, $\eta_m = 0.8$

Mean effective pressure $(P_m) = 0.35$ N/mm^2

For a given diesel engine, assume $\dfrac{l}{d} = 1.35$.

Now
$$BP = \frac{P_m l A n}{60 \times 1,000} \times \eta_m$$

$\Rightarrow$
$$5 = \frac{0.35 \times 1.35d \times \pi d^2 \times 0.8}{60 \times 1,000 \times 4} \times \frac{1,200}{2}$$

$\Rightarrow$
$$5 = 5.937 \times 10^{-3} d^3$$

$\Rightarrow$
$$d = 0.44 \text{ m} = 9,440 \text{ mm.}$$

also
$$l = 1.35d = 1.35 \times 9.44 = 12.744 \text{ m}$$

Thickness of cylinder head:

$$t_h = k_1 d \sqrt{\frac{P_{max}}{\sigma_{au}}}$$

$$\sigma_{au} = 40 \text{ N/mm}^2$$
$$k_1 = 0.35, \; P_{max} = 0.35 \text{ N/mm}^2$$

$$d = 9,440 \text{ mm}, \; t_h = 0.35 \times 9,440 \sqrt{\frac{0.35}{40}} = 309 \text{ mm} = 0.3 \text{ m}$$

2. A 4 stroke diesel engine has the following specifications:

Brake power: 12 kW

Speed: 1,500 rpm

Indicated *MEP*: 0.35 N/mm^2

Mechanical efficiency: 80%

Determine:

i. Bore and length of the cylinder

ii. Thickness of the cylinder head

iii. Size of the stud for cylinder head   [UPTU 2011]

SOLUTION: Given: $BP = 12$ kW $= 12000$ W, $n = \dfrac{N}{2} = \dfrac{1,500}{2} = 750$ rpm

$\qquad N = 1,500$ rpm, $P_m = 0.35$ N/mm^2, $\eta_m = 80\% = 0.8$

i. Bore and length of cylinder:

Let $\qquad d =$ Bore of the cylinder in mm

$$A = \text{Cross-sectional area of the cylinder} = \frac{\pi}{4}d^2 \text{ mm}^2$$

$$l = \text{length of the stroke} = \frac{1.5d}{1,000} \text{ m}$$

$$IP = \frac{BP}{\eta_m} = \frac{12,000}{0.8} = 15,000 \text{ W}$$

$$15,000 = \frac{P_m lAn}{60} = \frac{0.35 \times 1.5d \times \pi d^2 \times 750}{60 \times 1,000 \times 4}$$

Diameter (d) = 142.77 mm = 143 mm

Length (l) = $1.15l$ = 1.15 × 142.77 = 164.18 mm = 165 mm

ii. Thickness of the cylinder head:

Since the maximum pressure (P) in the engine cylinder is taken as 9 to 10 times of the mean effective pressure (P_m) therefore,

$$P = 9P_m = 9 \times 0.35 = 3.15 \text{ N/mm}^2.$$

Thickness of the cylinder head

$$t_h = d\sqrt{\frac{C_P}{\sigma_t}} = 143\sqrt{\frac{0.1 \times 3.15}{42}}$$

$$= 12.38 \text{ mm}$$

iii. Size of stud for the cylinder head

$$\frac{\pi}{4}d^2 P = \frac{\pi}{4}(143)^2 \, 3.15 = 50,590.912 \text{ N}$$

Let us consider no. of studs (n_s) = 6

$$n_s \times \frac{\pi}{4} \times (d_c)^2 \, \sigma_t = 6 \times \frac{\pi}{4} \times (0.84d)^2 \times 65 = 216.13d^2$$

$$= 50,590.912$$

$$d = 15.3 \text{ mm} = 16 \text{ mm}$$

The pitch circle diameter of studs (d_p) = $d + 3d$

$$= 143 + 3 \times 16 = 191 \text{ mm}$$

Minimum pitch of the studs $= 19\sqrt{d} = 19\sqrt{16} = 76$ mm

Maximum pitch of the studs $= 20.5\sqrt{d} = 20.5\sqrt{16} = 114$ mm.

6.5 DESIGN OF PISTON, PISTON RINGS AND GUDGEON PIN

Thickness of piston crown can be calculated by assuming it to be as flat plate of uniform thickness and fixed at the edges.

$$t = Cd\sqrt{\frac{P_{max}}{\sigma_t}}$$

where C = 0.43 to 0.5

d = Inner diameter of piston

t = Thickness of the crown

P_{max} = Maximum explosion pressure

σ_t = Allowable tensile strength

= 40 – 50 N/mm^2 (for cast iron) and 20–50 N/mm^2 (for aluminum)

Thickness can also be calculated on the basis of heat dissipation requirements instead of strength as:

$$t = \frac{d^2 q}{1,600k\left(T_c - T_e\right)}$$

where $\quad q$ = Heat flow from gases (J/sec m^2)

   = 32,000 to 1,28,000 for cast iron pistons of 4 stroke engines

   = 64,000 to 2,50,000 for aluminum alloy pistons

k = Heat conductivity of piston material

   = 460 J/sec m^2 °C/mm length for cast iron and

   = 1,600 J/sec m^2 °C/mm length for aluminum

T_c = 425 °C for cast iron and 260 °C for aluminum

$T_c - T_e$ = 220 °C for cast iron and 110 °C for aluminum

Piston section at *X–X* is weakened by oil return holes and is checked for compressive load due to gas force and tensile force due to inertia force.

For prevention of piston seizure during operation, diameters of the crown and skirt should have some diametral clearance between cylinder wall and the piston.

1. Clearance at crown (δ_c) = (0.004 – 0.008)d.
2. Clearance at skirt (δ_s) = (0.001 – 0.002)d.

Piston Rings

Basic functions of piston rings are:

i. Properly sealing combustion chamber without increasing friction.

ii. Help in transferring heat from piston to cylinder.

iii. Scrape oil off the cylinder wall on the down strokes, thereby preventing excessive oil from reaching the combustion chamber.

On this basis rings are of 2 types: Compression and oil control rings.

Radial thickness of cast iron piston ring is computed by the equation

$$t_{rad} = d\sqrt{\frac{3P_{rad}}{\sigma}}$$

where $\quad \sigma$ = Allowable stress of cast iron (80–100) N/mm^2

P_{rad} = Radial pressure on the ring (0.025 – 0.0 35) N/mm^2

Piston Pin

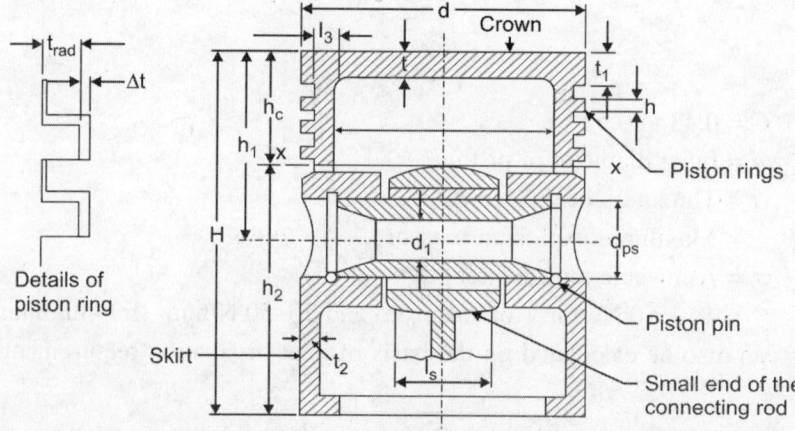

Fig. 6.8 (↑) Piston of IC engine and piston pin

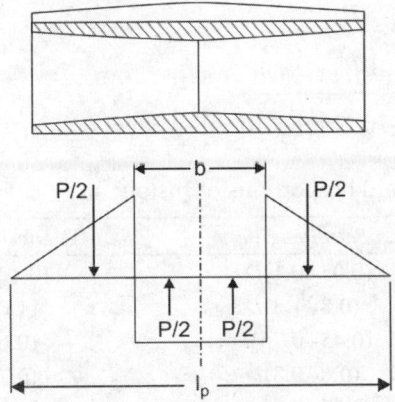

Fig. 6.9 Pressure distribution   **Fig. 6.10** Piston ring

Piston pin needs to be designed for the maximum gas load on the inertia force of the piston either of which being large. Maximum force is resisted by the bearing pressure between the piston pin and support length and depends upon the nature of connection between the piston pin and small end of connecting rod.

Let P = Maximum gas force

l_s = Length of the small end of the connecting

d_{ps} = Outer diameter of the piston pin

So bearing pressure $(P_{br}) = \dfrac{P}{d_{ps}l_s}$.

Permissible value of the bearing pressure is 8–16 N/mm² for petrol and diesel engines.

If piston pin is employed as floating on the bosses, the bearing pressure exerted is calculated as

$$P_{br} = \frac{P}{d_{ps}\left(l_p - b\right)}$$

where l_p is the length of piston pin

b is the distance between base ends

Ratio l_p/d_{ps} is usually between 1.5 and 2.5.

Assuming pressure distribution is replaced by the equivalent point load and the pressure distribution of piston pin as shown in above figure.

Maximum bending moment at center of the pin

$$M = \frac{P}{2} \times \frac{b}{4} - \frac{P}{2}\left[\frac{b}{2} + \frac{l_p - b}{6}\right]$$

Since piston pin is usually hollow, so section modulus

$$Z = \frac{\pi}{32} \frac{\left(d_{ps}^4 - d_i^4\right)}{d_{ps}}$$

here d_i = inner diameter of the piston pin

Bending stress $(\sigma_b) = M/Z$

Piston pin may fail in double shear near the bosses. Induced shear stress is

$$\tau = \frac{P}{2 \times \frac{\pi}{4}\left(d_{ps}^2 - d_i^2\right)}$$

The induced shear stress not exceed 60–70 N/mm² for alloy steel.

Table 6.1 Standard proportions of piston

Dimension	Petrol engine	Diesel engine
Piston crown thickness (t)	(0.05–0.1)D	(0.12–0.2)D
Piston height (H)	(0.8–1.3)D	(1.0–1.7)D
Height of piston (top part)(h_1)	(0.45–0.75)D	(0.6–1.0)D
Skirt length (h_2)	(0.6–0.8)D	(0.7–1.1)D
Top land (t_1)	(0.06–0.12)D	(0.11–0.2)D
Radial clearance of ring (Δt)	(0.7–0.95) mm	(0.7–0.95) mm
Length of pin (l_p)	(0.78–0.93)D	(0.8–0.93)D

WORKED EXAMPLES

1. Design on aluminum alloy piston for single acting four stroke engine from the following data:

> Piston diameter = 90 mm
> Speed = 1,500 rpm
> Length of the stroke = 99 mm
> Mean effective pressure = 0.7 N/mm²
> b_sf_c = 0.26 kg/kWh
> L/r ratio = 4

Heat conducted through the piston crown = 10% of heat generated during combustion
Calorific value of the fuel = 42 MJ/kg
Assume mechanical efficiency of the engine as 80%.

SOLUTION: Piston crown on the basis of strength

$$t = Cd\sqrt{\frac{P_{max}}{\sigma_t}}$$

Take $\qquad$ $C = 0.43$

Assume P_{max} (maximum explosion pressure) = 8 × Mean effective pressure
$$= 8 \times 0.7 = 5.6 \text{ N/mm}^2$$

σ_t (allowable stress) for aluminum alloy = 55 N/mm²

Hence $t = 0.43 \times 90\sqrt{\dfrac{56}{55}} = 12.34$, say 12 mm

On the basis of heat dissipation, $\quad t = \dfrac{d^2 q}{1,600k\left(T_c - T_e\right)}$

where $\qquad\qquad q =$ Heat flow from the gases J/sec m²

$$k = \text{Heat conductivity of piston material}$$
$$= 1,600 \text{ J/sec m}^2 \text{ °C /mm for aluminum alloy}$$
$$T_c - T_e = 110 \text{ °C for aluminum}$$

From thermodynamics:

Indicated power $(IP) = \dfrac{P_{im}LAN}{60 \times 2}$

$$= 0.7 \times 10^3 \times \frac{99}{1,000} \times \frac{\pi}{4} \times (0.090)^2 \times \frac{1,500}{60 \times 2}$$

$$= 0.7 \times 10^3 \times \frac{99}{1,000} \times \frac{\pi}{4} \times (0.090)^2 \times \frac{1,500}{120}$$

$$= 5.51, \text{ say 5 kW}$$

$$\eta_{mech} = \frac{BP}{IP}$$

$\Rightarrow$

$$BP = IP \times \eta_{mech} = 5 \times 0.8 = 4 \text{ kW}$$

Fuel consumption $= b_sf_c \times BP = 0.26 \times 4 = 1.04$ kg/hour

Heat supplied = Fuel consumption $\times CV$
(where CV = Calorific value)

$$= \frac{1.04 \times 42 \times 10^6}{3,600} = 12,133.3 \text{ J/sec}$$

Heat conducted through crown

$$= 10\% \text{ of heat supplied}$$
$$= 0.1 \times 12,133.3$$
$$= 1,213.33 \text{ J/sec}$$

Heat flow rate $(q) = \dfrac{\text{Heat conducted}}{\text{Cross-sectional area}}$

$$= \frac{1,213.33}{\dfrac{\pi}{4} \times (0.090)^2} = 190,646.68 \text{ J/sec m}^2$$

Thickness of the crown $(t) = \dfrac{(90)^2 \times 190,646.68}{160 \times 1,600 \times 110}$

$$= 5.48 \text{ mm}$$

Thus the acceptable value of the thickness of the crown $(t) = 12$ mm

Now we find radial thickness of the piston ring

$$\boxed{t_{rad} = d\sqrt{\frac{3P_{rad}}{\sigma}}}$$

where P_{rad} = Radial pressure on the ring

$$= 0.025 \text{ N/mm}^2 \text{ for four-stroke diesel engine}$$

σ = Allowable stress for cast iron

$$= 85 \text{ N/mm}^2$$

Thus
$$t_{rad} = 90\sqrt{\frac{3 \times 0.025}{85}}$$
$$= 2.67 \text{ mm, say 4 mm.}$$

Width of the ring (h) = $(0.7–1.0)\, t_{rad}$
$$= 0.8 \times 0.4 = 3.2 \text{ mm}$$

Number of rings $(i) = \dfrac{d}{10h} = \dfrac{90}{10 \times 32} = 2.8$

Shape of the piston ring:

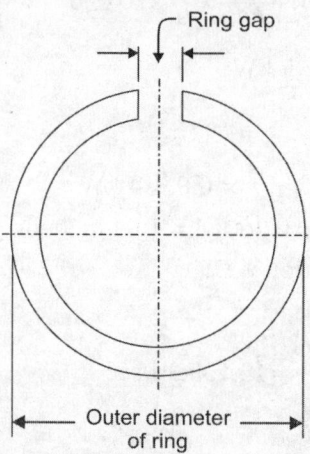

Ring gap

Outer diameter
of ring

Considering 3 compression rings and one oil ring
Distance between the first ring groove and top
Surface, i.e. top land (t_1) = $(0.08 \text{ to } 0.2)d$
$$= 0.08d = 0.08 \times 90 = 7.2 \text{ mm say 7mm}$$

Thickness of the piston crown wall
$$t_3 = (0.05 \text{ to } 0.1)d = 0.1d = 9 \text{ mm}$$

Piston inner diameter
$$d_i = d - 2(t_3 + t_{rad} + \Delta t)$$

Here Δt = Radial clearance of ring
$$= 0.7 \text{ to } 1.1 \text{ mm} = 0.8 \text{ mm}$$

Therefore, d_i = $90 - 2(9 + 4 + 0.8)$
$$= 90 - 27.6 = 62.4 \text{ say 60 mm}$$

Thickness of the spirit wall (t_2) = $(2 \text{ to } 5) \text{ mm} = 4 \text{ mm}$

Piston Pin

Induced bearing pressure $(P_{br}) = \dfrac{P}{d_{ps} \times l_s}$

Gas force = $P \times \dfrac{\pi}{4} \times (90)^2 \times 5.6 = 35.64 \text{ kN}$

Assuming $l_s/d_{ps} = 1.5$

and permissible bearing pressure (P_{br}) = 20 N/mm^2

Thus $$\frac{P}{P_{br}} = d_{ps} \times 1.5 d_{ps}$$

or $$d_{ps}^2 = \frac{P}{1.5 P_{br}}$$

or $$d_{ps} = \sqrt{\frac{P}{1.5 P_{br}}} = 34.46 \text{ mm, say 35 mm}$$

and inside diameter $(d_i) = 0.6 \times d_{ps} = 21$ mm

Empirically, $l_p = (0.80 \text{ to } 0.93)\, d$

Let us adopt, $l_p = 0.9d = 0.9 \times 90 = 81$ mm, say 80 mm

$b = $ Length of small end of the connecting rod + clearance

$= 1.5 d_{ps} + 4$

$= 1.5 \times 35 + 4 = 56.5$ mm, say 56 mm

Bending moment $$(M) = \frac{p \times b}{4} - \frac{P}{2}\left[\frac{b}{2} + \frac{l_p - b}{6}\right]$$

$$= \frac{35,640}{4} - \frac{35,640}{2}\left[\frac{56}{2} + \frac{80 - 56}{6}\right]$$

$$= 69,580 \text{ N-mm}$$

Section modulus $$(Z) = \frac{\pi}{32}\left[\frac{(35)^4 - (21)^4}{35}\right] = 3,663.7 \text{ mm}^2$$

Bending stress $$(\sigma_p) = \frac{M}{2} = \frac{69,580}{3,663.7} = 18.99 \text{ N/mm}^2 \text{ (which is quite reasonable)}$$

Piston may fail in double shear

Induced shear stress $$(\tau) = \frac{2P}{\pi\left(d_{ps}^2 - d_i^2\right)} = \frac{2 \times 35,640}{\pi\left[\left(35^2 - 21^2\right)\right]}$$

$$= \frac{498,960}{17,248} = 28.92 \text{ N/mm}^2 \text{ (which is quite reasonable)}$$

PREVIOUS YEAR UNIVERSITY QUESTIONS

1. Design a piston for a single acting 4 stroke engine for the following specifications: Cylinder bore = 0.30 m, stroke length = 0.375 m, maximum gas pressure = 8 MPa, brake mean effective pressure = 1.15 MPa, fuel consumption = 0.22 kg /kW/hour, speed = 500 rpm. Assume suitable data. (UPTU 2008)

SOLUTION: Given: $d = 0.30$ m, $l = 0.375$ m

$P = 8$ MPa $= 8$ N/mm^2, $P_m = 1.15$ MPa $= 1.15$ N/mm^2

$m = 0.22$ kg/kW/hour, $N = 500$ rpm

Piston head or crown:

$$T_H = \sqrt{\frac{3 P_m d^2}{16 \sigma_t}} = \sqrt{\frac{3 \times 1.15 \times (30)^2}{16 \times 38}} = 23 \text{ mm}$$

(assuming $\sigma_t = 38$ MPa for CI)

Since the engine has four stroke, so

$$n = \frac{N}{2} = \frac{500}{2} = 250 \text{ rpm}$$

Cross-sectional area of cylinder $= \dfrac{\pi}{4}d^2 = \dfrac{\pi}{4} \times (0.30)^2 = 0.0707\,\text{m}^2 = 70,685\ \text{mm}^2$

$$BP\ (\text{brake power}) = \frac{P_m LAN}{60}$$

$$= \frac{1.15 \times 375 \times 70,685 \times 250}{60 \times 1,000}$$

$$= 127,012\ \text{W} = 127.012\ \text{kW}$$

Heat flowing through the piston head

Taking $C = 0.05$

$$HC_V = 42 \times 10^3\ \text{W/kg}$$

$$H = C \times HC_V \times m \times BP$$

$$= \left(0.05 \times 42 \times 10^3 \times \frac{0.22}{3,600} \times 127.012\right)$$

$$= 0.0162998 \times 10^3\ \text{W}$$

$$= 16.2998\ \text{W} = 0.16\ \text{kW}$$

Thickness of the piston head

$$t_h = \frac{H}{12.56k\,(T_c - T_e)}$$

$$= \frac{0.0162998}{(12.56 \times 46.6 \times 220)}$$

$$= 13\ \text{mm}$$

Radial ribs:

Thickness of ribs $= t_h/3 = 13/3 = 5$ mm.

Piston rings:

Radial thickness of piston rings

$$t_1 = d\sqrt{\frac{3P_W}{\sigma_E}} = 300\sqrt{\frac{3 \times 0.35}{90}}$$

$$= 10.246\ \text{mm.}$$

Axial thickness of piston rings $(t_2) = 0.7t_1 = 0.7 \times 10.24 = 7.168$ mm.

Minimum axial thickness of the piston ring

$$t_2 = \frac{d}{10n} = \frac{300}{10 \times 4} = 7.5\ \text{mm}$$

Distance from the top of the piston to the first ring groove

$$b_1 = t_h \text{ to } 1.2\ t_h = 23 \text{ to } 27.6\ \text{mm}$$

Width of other ring $(b_2) = 0.75\ t_2$ to t_2

$$= 0.75 \times 7.168 \text{ to } 7.168$$

$$= 5.376 \text{ to } 7.168\ \text{mm}$$

∴ $\qquad\qquad b_1 = 28$ mm, $b_2 = 7$ mm

Gap between the free ends of the ring

$$C_{11} = 3.5\ t_1 \text{ to } 4t_1 = 35.84 \text{ to } 40\ \text{mm.}$$

Gap when the ring is in the cylinder

$$C_{12} = 0.002d \text{ to } 0.004d = 0.002 \times 300 \text{ to } 0.004 \times 300$$

$$= 0.6 \text{ to } 1.2\ \text{mm}$$

Piston Barrel

$$b = t_1 + 0.4 = 10.24 + 0.4 = 10.64 \text{ mm}$$
$$t_3 = (0.03d + b + 0.4) \text{ mm}$$
$$= (0.03 \times 300 + 10.64 + 4.5) = 24.14 \text{ mm}$$
$$t_4 = 0.24t_3 \text{ to } 0.35t_3 = 0.25 \times 24.14 \text{ to } 0.35 \times 24.14$$
$$= 6.035 \text{ to } 8.449 \text{ mm}$$

Piston Skirt

l = Length of the skirt in mm.

Maximum side thrust on the cylinder due to gas pressure (P)

$$R = \mu \times \frac{\pi d^2}{4} \times P = 0.1 \times \pi \times \frac{(300)^2}{4}$$

$$= 7,068.58 \text{ N}$$

Side thrust on the cylinder due to gas pressure (P_b)

$$R = P_b \times d \times l = 0.45 \times 300 \times l = 135l$$
$$135l = 7,068.58, \quad l = 50.36 \text{ mm}$$

Total length of piston (L) = Length of the skirt + Length of the ring section + Top band

$$= l + (4t_2 + 3b_2) + b_1$$
$$= 52.36 + (4 \times 7.5 + 3 \times 7) + 2.8$$
$$= 131.36 \text{ mm}$$

2. Design an aluminum alloy piston for a single acting four stroke engine for the following specifications: (UPTU 2010)

Cylinder bore = 0.30 m

Stroke = 0.375 m

Maximum gas pressure = 8 MPa

Brake mean effective pressure = 1.15 MPa

Fuel consumption = 0.22 kg/kW/hour

Speed = 500 rpm

SOLUTION: Piston material: Al

Given: Single acting, 4 stroke $d = 0.3$ m $= 0.3 \times 1,000 = 300$ mm

$$n = \frac{500}{2} = 250 \text{ rpm}$$

$$l = 0.375 \text{ m} = 375 \text{ mm}$$

$$P_{max} = 8 \text{ N/mm}^2, \ P_{mep} = 1.15 \text{ MPa}, \ b_s f_c = 0.22 \text{ kg/kW/hour}$$

Step 1

a. Thickness of piston head based on strength

$$t_{ph} = 0.433D \sqrt{\frac{P_{max}}{\sigma_{min}}} = 0.433 \times 300 \times \sqrt{\frac{8}{50.70}}$$

$$= 51.96 \text{ mm (max) and } 43.91 \text{ mm (min)}$$

$$t_{ph} = \frac{45}{50} \text{ mm}$$

b. Thickness at piston head based on heat dissipation

$$P_{bm} = 1.15 \text{ MPa} = 1.15 \times 10^6 \text{ N/m}^2$$

$$l = 0.375 \text{ m}, \quad A = \frac{\pi}{4}d^2 = \frac{\pi}{4}(0.3)^2 = 0.07068 \text{ m}^2$$

$$BP = \frac{P_m lAn}{60 \times 1,000} = 127 \text{ kW}$$

$$b_s f_c = 0.22 \text{ kg/kW/hour}$$

$$= \frac{0.22}{3,600} = 6.11 \times 10^5 \text{ kg/kW/sec}$$

Assuming *HCV* of fuel = 40,000 to 50,000 kJ/kg

For Al, $k = 175$ W/m °C at 75 °C, $x = 0.05$

$$t_{ph} = \frac{x(b_s f_c)(BP)(HCV)}{4\pi kAT}$$

$$= \frac{0.05 \times 6.11 \times 10^5 \times 127 \times 10^3 \times (40,000)}{4\pi \times (175) \times (75)} = 0.094 \text{ m}$$

$$= 94 \text{ mm}$$

Step 2

Thickness of piston barrel under the ring $(t_{pb}) = 94$ mm

Step 3

Design of piston ring radial thickness $(t_r) = d\sqrt{\dfrac{3P_w}{\sigma_{b_r}}} = 10.39$ mm

Axial width of piston rings (b_r) = 0.75 to 1.0

$$= 0.75 \times 10.39 \text{ to } 10.39$$

$$= 8\text{–}9 \text{ mm}$$

Number of piston rings = $\dfrac{d}{102}$ = 3 to 4

Width between piston ring grooves $(b_1) = 0.75b_r$ to b_r

$$= 0.75 \times 10 \text{ to } 10 = 7.5\text{–}10 \text{ mm} = 8 \text{ mm}$$

Width at top land $(b) = 1t_r$ to $1.25t_r = 94$ to 1.25×94

$$= 94 \text{ to } 117.5 \text{ mm} \approx 100 \text{ mm}.$$

Gap between free ring ends

$$G_r = 3t_r \text{ to } 4t_r = 3 \times 10.39 \text{ to } 4 \times 10.39$$

$$= 35 \text{ mm}$$

6.6 DESIGN OF CONNECTING ROD

Stresses set up in the connecting rod are due to:

1. Gas pressure and the piston inertia
2. Friction of piston rings and piston
3. Inertia of connecting rod
4. Frictional force in big and small end bearings

Maximum force due to gas pressure

$$P_{gas} = P_{max} \times \text{Area of cross-section}$$

Maximum explosion pressure is about 8 to 10 times of the indicated mean effective pressure.

$$P_{inertia} = -mr\omega^2 \left[\cos\theta + \frac{\cos 2\theta}{n} \right]$$

where m = Mass of the piston + $\frac{1}{3}$ times mass of the connecting rod

ω = Angular speed (rad/sec)

θ = Crank angle

n = Ratio of length of the connecting rod to crank radius (l/r)

Net force acting on the piston will be the algebraic sum of the gas force and the inertia force

$$P = P_{gas} + P_{inertia}$$

Net force acting on the piston is transferred to the connecting rod.

Direct force acting on connecting rod $(Q) = P/\cos\phi$

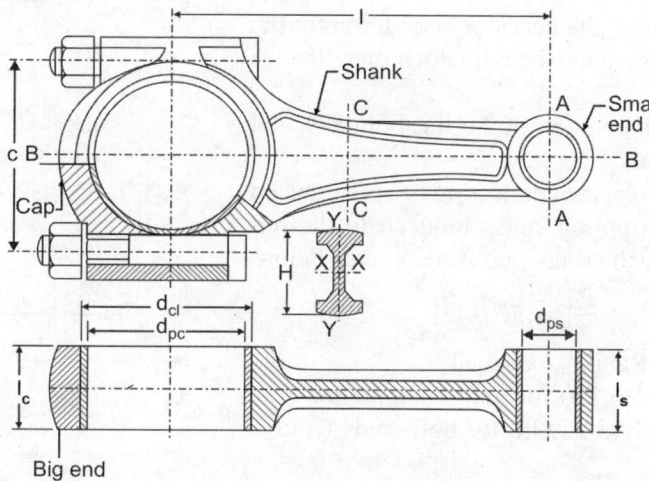

Fig. 6.11 Connecting rod

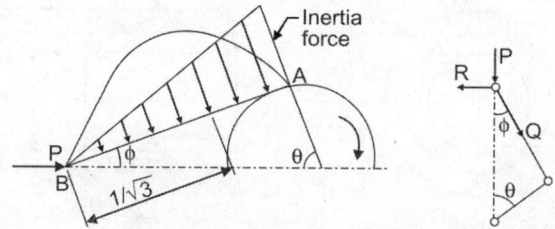

Fig. 6.12 Forces on connecting rod

where ϕ is the obliquity or angularity that can be determined by the following relation:

$$\sin\phi = \frac{\sin\theta}{n}$$

Bending loads: Small end of the connecting rod has a motion of pure translation and the big end has a rotary motion while all the intermediate points on the rod move in the elliptical orbits. Transverse or lateral oscillations of the rod section result in inertia bending forces all along the length of the rod. This action is termed as whipping action and the stress induced in rod is called whipping stress.

$$\text{Inertia force} \qquad \boxed{F_{inertia} = \frac{\rho A \omega^2 rl}{2}}$$

Maximum bending moment acts at a distance $l/\sqrt{3}$ from the piston pin end.

$$\boxed{M_{max} = \frac{2F_{inertia}l}{9\sqrt{3}} = 0.1283\, F_{inertia}l}$$

Value of the crank angle θ at which the bending moment is maximum is given by

$$\boxed{\theta = 90° - \frac{3{,}500}{(n+7.82)^2}}$$

Most suitable section for the connecting rod shank is the I-section (shown in the figure).

For an I-section rod, the necessary condition for the rod to be equally resistant to buckling in either plane is $I_{xx} = 4I_{yy}$

where I_{xx}, I_{yy} are the moments of inertia about X–X and Y–Y axes respectively.

Due to the gas force, the connecting rod is subjected to buckling. The crippling stress induced in the rod can be computed by the Rankine formula given below.

$$\sigma_{cr} = \frac{P}{A}\left[1 + a(l/k)^2\right]$$

where
a = Rankine constant
= 1/6,250 for both ends hinged
= 1.95/25,000 for both ends fixed

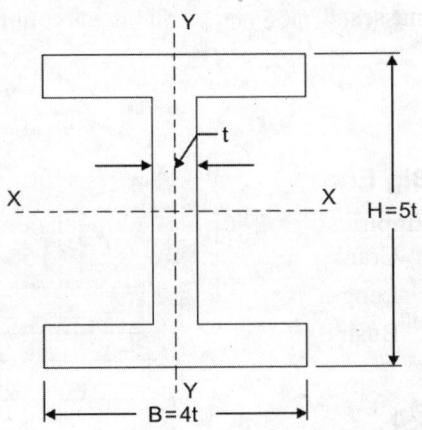

Fig. 6.13 Section of connecting rod shank

Small End

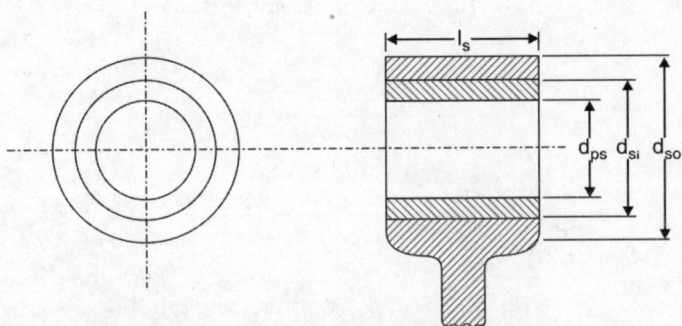

Fig. 6.14 Small end of connecting rod

Dimensions of small end may be obtained by the following empirical relations:

Inner diameter of the small end $(d_{si}) = (1.1 - 1.25)\, d_{ps}$

Outer diameter of the small end $(d_{so}) = (1.25 - 1.65)d_{ps}$

Length of the small end $(l_s) = (0.3 - 0.45)D$

where d_{ps} = Outer diameter of the piston pin

d = Cylinder bore

For small end of the connecting rod:

Bearing failure of pin

$$P_{br} = \frac{P}{l_s d_{ps}} \leq P_{allowable}$$

P_{br} = Bearing pressure

Bending failure of pin

$$\sigma_b = \frac{M}{Z} \leq \sigma_{allowable}$$

This failure is due to the gas force.

Upper part of the small end is subjected to tensile stress due to inertia of the reciprocatory masses. This inertia force is maximum when the crank rotation angle is 0°.

$$\boxed{\sigma_t = \frac{P_{inertia}}{(d_{so} - d_{si})l_s} \leq \sigma_{allowable}}$$

Big End

Empirical relations for the dimensions of the 'big end' are as follows:

Crank pin diameter (d_{pc}) $= (0.55 - 0.75)d$

Length of the big end (l_c) $= (0.45 - 1.0)d_{pc}$

Bush thickness (t_{bush}) $= (0.03 - 0.1)d_{pc}$

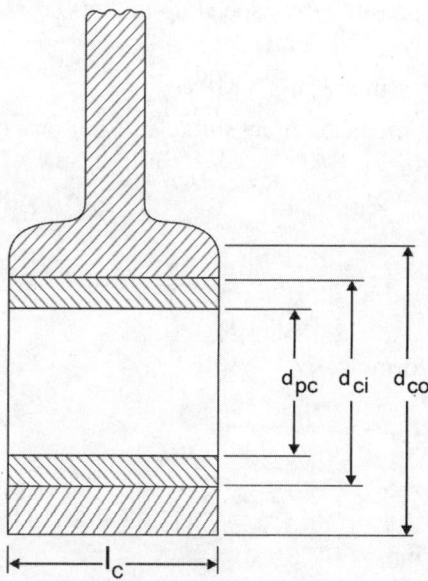

Fig. 6.15 Big end of connecting rod

Big end of the connecting rod is usually made in the halves, which are fastened by two bolts. The maximum force on the bolts and on the big end will be the inertia force at *TDC* of the suction stroke.

Total force on the bolts

$$P_{bolt} = F_{initial} + \frac{KP_{inertia}}{N}$$

Here, $F_{initial}$ = Initial tightening force
    k = Gasket factor
    N = Number of bolts

Cap of Big End

Cap of big end is designed as a beam supported at the bolt center and may be assumed to be loaded with concentrated load.

The thickness of the cap is computed by the relation

$$t_c = \sqrt{\frac{P_{inertia} c}{l_c \sigma_y}}$$

where c is the distance between the bolt centers.

WORKED EXAMPLES

1. Design a connecting rod for a petrol engine from the following data:

   Diameter of piston: 110 mm
   Mass of reciprocating parts: 2 kg
   Length of connecting rod: 325 mm
   Stroke: 150 mm
   Speed: 1,500 rpm with possible over speed up to 2,500 rpm
   Compression ratio: 4:1
   Maximum explosion pressure (P_e): 2.5 MPa

SOLUTION: Considering I-section as the most suitable, usual proportions are:

   Width of flange $(B) = 4t_o$
   Height of I-section $(H) = 5t_o$
   Web thickness $= t_o$

Maximum gas force $(P_{gas}) = \dfrac{\pi}{4} D^2 \times P_{max} = \dfrac{\pi}{4} \times (110)^4 \times 25$

$$= 23{,}758.3 \text{ N}$$

According to Rankine formula

$$F_{cr} = \frac{\sigma_{cr} A}{1 + a(l/k)^2}$$

Taking factor of safety = 5

$$F_{cr} = P_{gas} \times 5 = 23{,}758.3 \times 5 = 118{,}791 \text{ N}$$

Assuming $\sigma_{cr} = 460 \text{ N/mm}^2$

$$= \frac{1}{6{,}250} \text{ (for both ends hinged)}$$

$$K_{xx}^2 = I_{xx}/A = 31.8 t^2$$
$$A = 11 t^2$$

Thus
$$118,791 = \frac{460 \times 10^6 \times 11t^2}{1 + \frac{1}{6,250}\left(\frac{0.325^2}{3.18t^2}\right)} \text{ or } t = 5\,\text{mm}$$

Let the height of I-section near the crank pin = 1.2 × 25 = 30 mm
and height of I-section near the piston pin
$$= 0.8 \times 25 = 20 \text{ mm}$$
Considering bearing failure of the pin

$$P_{br} = \frac{P_g}{l_s d_{ps}}$$

Assume allowable bearing pressure
$$P_{br} = 15 \text{ N/mm}^2$$

and
$$l_s/d_{ps} = 2 \Rightarrow d_{ps} = \sqrt{\frac{P_{gas}}{2P_{br}}} = \sqrt{\frac{23,758.3}{2 \times 15}} = 28.14 \text{ mm, say 28 mm}$$

Thus length of the small end
$$l_s = 2d_{ps} = 2 \times 28 = 56 \text{ mm}$$
Inner diameter of the small end (d_{si}) $= 1.15 d_{ps}$
$$= 1.15 \times 28$$
$$= 32.2 \text{ mm, say 32 mm}$$
Outer diameter of the small end
$$d_{so} = 1.4 d_{ps}$$
$$= 1.4 \times 28 = 39.2 \text{ mm}$$
$$= 39 \text{ mm}$$

A babbitt or bronze metal bush of 2 mm is inserted by considering bearing failure of crank and assuming empirical relations.
Diameter of the crank pin $\quad (d_{pc}) = (0.55 - 0.75)d$
$$= 0.6 \times D = 0.6 \times 110$$
$$= 66 \text{ mm}$$
Length of the crank pin $\quad (l_c) = 1.0 d_{pc} = 66 \text{ mm}$

Bearing pressure
$$P_{br} = \frac{P_{gas}}{l_c d_{pc}} = \frac{23,758.3}{66 \times 66}$$

$$= 5.45 \text{ N/mm}^2$$
which is safe according to design considerations.
Thickness of bush $\quad (t_{bush}) = (0.03 - 0.1)d_{pc}$
$$= 0.05 d_{pc}$$
$$= 0.05 \times 66 = 3.3 \text{ mm, say 3 mm}$$

The maximum force on the bolt at the big end cap will be the inertia force at *TDC* of the section side.

$$P_{inertia} = m\omega^2 r\left[\cos\theta + \frac{\cos 2\theta}{n}\right]$$

where
$$n = \frac{l}{r} = \frac{325}{70} = 4.6, \ \theta = 0°$$

$$\omega = \frac{2\pi \times N}{60} = \frac{2\pi \times 500}{60} = 261.79 \text{ rad/sec}$$

$$m = 2 \text{ kg}$$

Thus

$$P_{inertia} = 2 \times (261.79)^2 \times \frac{70}{1,000}\left(1 + \frac{1}{4.6}\right)$$

$$= 11,680.5 \text{ N}$$

Let us assume number of bolts $(N) = 2$

Tensile strength $= 600 \text{ N/mm}^2$, $F_s = 5$

Allowable strength $(\sigma) = \dfrac{600}{5} = 120 \text{ N/mm}^2$

Initial tightening force

$$F_{initial} = 2.5 \times \frac{P_{inertia}}{N} = \frac{2.5 \times 11,680.5}{2}$$

$$= 14,600.62 \text{ N}$$

Total force $\quad (P_{bolt}) = F_{initial} + \dfrac{kP_{inertia}}{2}$

where k = Gasket factor = 0.2 (for hard gasket)

Thus

$$P_{bolt} = 14,600.62 + 0.2 \times \frac{11,680.5}{2}$$

$$= 15,768.67 \text{ N}$$

$$\text{Resisting area } (A) = \frac{P_{bolt}}{\sigma_t} = \frac{15,786.6}{120} = 131.40 \text{ mm}^2$$

As per IS: 4218–1967, size of the bolt may be found.

The suitable size is M14 × 1.5 mm

Now the cap may be assumed to be loaded at the center by a concentrated load.

Bending moment $(M) = \dfrac{P_{inertia} \times c}{6}$

where $\qquad c$ = Distance between bolt center, empirically

$\qquad c = (1.3 \text{ to } 1.75)d_{pc}$

Take $c = 95$ mm which is within the range.

Bending moment $(M) = \dfrac{P_{inertia} \times c}{6} = \dfrac{11,680.5 \times 95}{6}$

$$= 184,991.25 \text{ N-mm}$$

Assuming allowable stress of cap material as 100 N/mm^2

Section modulus $(Z) = \dfrac{1}{6}l_c t_c^2$

$$= \frac{1}{6} \times 66 \times t_c^2$$

$$= 11 t_c^2$$

Bending stress $(\sigma_b) = \dfrac{M}{2} \leq 1\ \sigma_{allowable}$

$\dfrac{184{,}991.25}{11 t_c^2} = 100$ or $t_c = 12.96$ mm (say 13 mm)

2. Following data refers to a single cylinder 4-stroke petrol engine.

    Cylinder diameter = 25 cm

    Stroke = 40 cm

    Maximum explosion pressure = 2.4 MPa at *TDC*

    Gas pressure at maximum torque

    Position when crank angle is 40° = 0.9 MPa

    Weight of flywheel = 16 kN

    Total tension in bolts on tight and slack side = 3.5 kN

    Length of connecting rod = 95 cm

    Speed = 1,000 rpm

  Design a connecting rod having I-section for above engine.

SOLUTION: Cylinder diameter/piston diameter (d) = 250 mm

    Stroke (S) = 400 m = $2r$

    Maximum explosive pressure = 2.4 MPa

    Crank angle (θ) = 40°

Gas pressure at maximum torque position

    $P' = 0.9$ MPa

    Weight of flywheel = 16 kN

    Tension in belt = 3.5 kN

    Length of CR = 950 mm

    Speed = 1,000 rpm

Dimensions of I-section of Connecting Rod

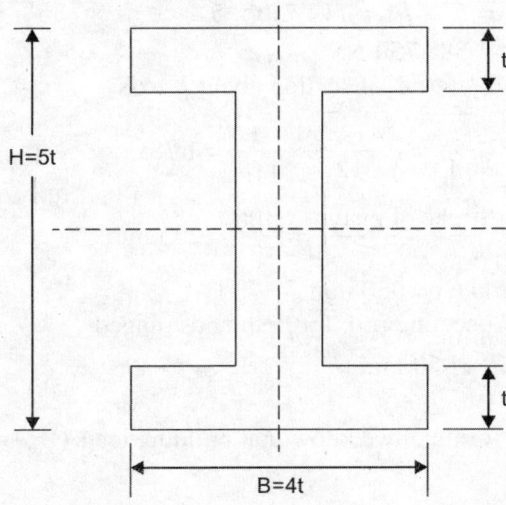

Let us consider an I-section of connecting rod as shown in figure: (with proportions)

Width of section $(B) = 4t$

Depth or height of section $(H) = 5t$

First of all, let us find whether section chosen is satisfactory or not.

Connecting rod is considered like both ends hinged. For buckling about X-axis and both ends fixed for buckling about Y-axis. The connecting rod should be equally strong in buckling about both the axes. In order to have a connecting rod, equally strong about both the axes.

$$I_{xx} = 4I_{yy}$$

I_{xx} = Moments of inertia of section about X-axis

I_{yy} = Moments of inertia of section about Y-axis

In actual practice, I_{xx} is kept less than $4I_{yy}$ and it is taken between 3 to 3.5 and connecting rod is designed for buckling about X-axis.

For section as shown in figure of I-section (previous page)

Area of the section

$$A = 2(4 \times t \times t) + 3t \times t = 11t^2$$

$$I_{xx} = \frac{1}{12}[4t(5t)^3 - 3t(3t)^3] = \frac{419}{12}t^4$$

$$I_{yy} = 2 \times \frac{1}{12}t(4t)^3 + \frac{1}{12} \times 3t \times t^3 = \frac{131}{12}t^4$$

$$\frac{I_{xx}}{I_{yy}} = \frac{419}{12}r^4 \times \frac{12}{131}t^4 = 3.2$$

Since $\dfrac{I_{xx}}{I_{yy}} = 3.2$, therefore, section chosen is quite satisfactory.

Force of connecting rod (F_c) is equal to maximum force on the piston (F_t) due to gas pressure.

Therefore, $F_c = F_L = \dfrac{\pi d^2}{4} \times P = \dfrac{\pi \times (250)^2}{4} \times 2.4 = 117,750$ N

Connecting rod is designed for buckling about X-axis

Assuming both the ends hinged

and factor of safety equal to 'S'

So buckling load $(W_b) = F_c \times FS = 117,750 \times 5$

or $\qquad W_b = 588,750$ N

We know that radius of gyration of section about X-axis

$$K_{xx} = \sqrt{\frac{I_{xx}}{A}} = \sqrt{\frac{419t^4}{12} \times \frac{1}{11t^2}} = 1.78t$$

Length of crank (r) $\quad = \dfrac{\text{Stroke of piston}}{2} = \dfrac{400}{2} = 200$ mm

Length of connecting rod $(l) = 950$ mm

Equivalent length of connecting rod, for both ends hinged

$$L = l = 950 \text{ mm}$$

According to Rankine formula, we know that building load $(W_b) = \dfrac{\sigma_c A}{1 + a\left(\dfrac{L}{K_{xx}}\right)^2}$

Selecting forged steel, material for connecting rod, where compressive yield stress σ_c is about 400 N/mm^2.

$$a = \frac{1}{7,500}$$

$$\therefore \quad 588,750 = \frac{400 \times 11t^2}{1 + \frac{1}{7,500}\left(\frac{380}{1.78t}\right)^2}$$

So $\quad 1,472 = \frac{11t^4}{t^2 + 6.1} \quad \Rightarrow \quad t^4 - 134t^2 - 813 = 0 \qquad$ (keeping $t^2 = T$)

$\Rightarrow \quad T^2 - 134T - 813 = 0$

$\Rightarrow \quad T = t^2 = 139.73$

or $\quad t = 11.82$, say 12 mm

Thus dimensions of I-section of connecting rod are thickness of flange and web of section

$t = 12$ mm

Width of section $(B) = 4t = 4 \times 12 = 48$ mm

Depth or height of section $(H) = 5t = 5 \times 12 = 60$ mm

These dimensions are at the middle of connecting rod. The width (B) is kept constant throughout the length of the rod, but the depth (H) varies, the depth near big end or crank end is kept as $1.1H$ to $1.25H$ and depth near small end or piston end is kept as $0.75H$ to $0.9H$

Taking depth near big end: $H_1 = 1.2H = 1.2 \times 60 = 72$ mm

Depth near small end

$H_2 = 0.85H = 0.85 \times 60$

$H_2 = 51$ mm

Dimension of section near big end = 72 × 48 mm

Dimensions of section near small end = 51 mm × 48 mm

Since connecting rod is made by forging, therefore, sharp corners of I-section are rounded as shown in the figure below.

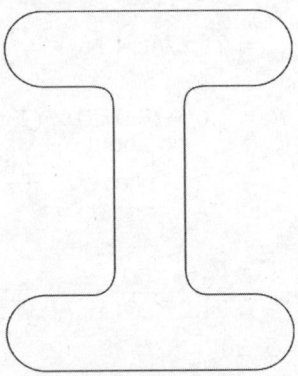

PREVIOUS YEAR UNIVERSITY QUESTIONS

1. A connecting rod is required to be designed for a high speed, 4 stroke IC engine:

(UPTU 2002)

Following data are available:

Diameter of piston = 88 mm

Mass of reciprocating parts = 1.6 kg

Length of connecting rod (center to center) = 300 mm

Stroke = 125 mm

RPM = 2,200 (when developing 50 kW)

Compressing ratio = 6.8:1 (approx)

Probable maximum explosion pressure (assumed shortly after dead center, say at about 3° angle) = 3.5 N/mm²

SOLUTION: Given:

Diameter of piston (d) = 88 mm

Mass of reciprocating parts (M) = 1.6 kg

Length of connecting rod (l) = 300 mm

Stroke length (L) = 150 mm

Speed (RPM) (N) = 2,200

Power developed (P) = 50 kW

Compression ratio = 6.8:1

Maximum explosive pressure (P_e) = 3.5 N/mm²

P_e = 3.5 MPa

At $\theta = 3°$, crank radius $(r) = \dfrac{L}{2} = \dfrac{150}{2} = 75$ mm

$$n = \frac{l}{r} = \frac{300}{75} = 4$$

Angular speed $\qquad \omega = \left(\dfrac{2\pi \times 2,200}{60}\right)$

or $\qquad\qquad\qquad \omega = 230.26$ rad/sec

Maximum force on piston due to pressure

$$F_p = \frac{\pi}{4}d^2 P_e = \frac{\pi}{4}(0.088)^2 \times 3.5 \times 10^6 \text{ N}$$

$$F_p = 21,276.64 \text{ N}$$

Inertia force of reciprocating parts

$$F_i = 1.6 \times (0.075) \times (230.26)^2$$

$$\left[\cos(3) + \frac{\cos(2 \times 3)}{4}\right] = 7,934.5 \text{N}$$

Net force on gudgeon pin at $\theta = 3°$

$$F = F_p - F_i$$

$$= 21,276.64 - 7,934.5 = 13,342.15 \text{ N}$$

Force in the connecting rod at $\theta = 3°$

$$F_c = \frac{F}{\cos\theta} = \frac{F}{\sqrt{1 - \dfrac{\sin^2\theta}{n^2}}} = \frac{13,342.15}{\sqrt{1 - \dfrac{(\sin 3)^2}{16}}}$$

$$F_c = 13,343.2$$

For the connecting rod to be equally strong about X- and Y-axis: $4k_{yy}^2 = 4k_{xx}^2$

Cross-section satisfying the condition is I-section shown below:

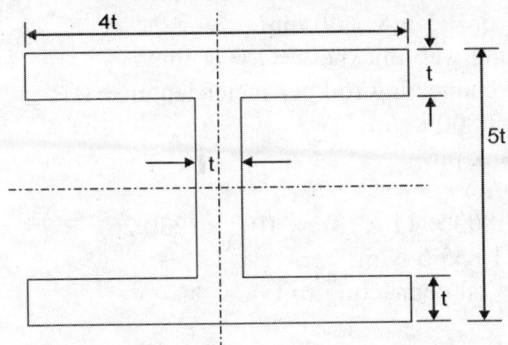

$$I_{xx} = \frac{1}{12}\left[4t \times (5t)^3 - 3t(3t)^3\right]$$

$$= \frac{t^4}{12}(500 - 81) = 34.917t^4$$

$\Rightarrow$ $$I_{yy} = \frac{1}{12}\left[2 \times t \times (4t)^3 - 3t \times t^3\right]$$

$$= \frac{t^4}{12}(128 + 3) = 10.917t^4$$

$\therefore$ $$\frac{I_{xx}}{I_{yy}} = 3.2$$

Area of cross-section $(A) = 4t \times 5t - 3t \times 3t$

or $A = 11t^2$, $$K_{xx}^2 = \frac{I_{xx}}{A} = \frac{34.917t^4}{11t^2} = 3.174t^2$$

Using Rankine-Gorden formula

$$P_c = \frac{\sigma_c A}{1 + a\left(\dfrac{l_c}{K_{xx}}\right)^2}$$

Taking $\sigma_c = 130$ MPa (for connecting rod)

$$a = \frac{1}{7,500} \text{ and factor of safety} = 5$$

We have $$F_c \times 5 = \frac{130 \times 10^6 \times 11t^2}{1 + \dfrac{1}{7,500}\left(\dfrac{1}{K_{xx}}\right)^2}$$

$$13,343.2 \times 5 = \frac{130 \times 10^6 \times 11t^2}{1 + \dfrac{1}{7,500}\left(\dfrac{0.300t^2}{3.174t^2}\right)}$$

So $t^2 = 46.65 \times 10^{-6}$

$\Rightarrow$ $t = 6.8$ mm or $t \approx 8$ mm

Width of section = $4t = 4 \times 8 = 32$ mm

Depth = $5t = 5 \times 8 = 40$ mm

Flange and web thickness = $t = 8$ mm

Mass of connecting rod per meter length = AP

Let $P = 7,800$ kg/m^3

Maximum force in crank pin

$F_{max} = \rho A \omega^2 r$

$F_{max} = 7,800 \times 11 \times (8)^2 \times 10^{-6} \times (230.26)^2 \times 75 \times 10^{-3}$

$F_{max} = 21,835.6$ N/m

Resultant normal force on connecting rod

$$F_n = \frac{1}{2} F_{max} l$$

$$= \frac{1}{2} \times 21,835.6 \times 300 \times 10^{-3}$$

$$F_n = 3,275.34 \text{ N}$$

Maximum bending moment

$$M_{max} = \frac{2 F_n l}{9\sqrt{3}} = \frac{2 \times 3,275.34 \times 300 \times 10^{-3}}{9\sqrt{3}}$$

$$= 126.07 \text{ N-m}$$

$$\sigma_{max} = \frac{M_{max}}{Z_{xx}} \Rightarrow Z_{xx} = \frac{I_{xx}}{2.5t}$$

$$= \frac{34.917 t^4}{2.5t}$$

$\Rightarrow$
$$Z_{xx} = 13.97 \times (8 \times 10^{-3})^3$$

$$= 7.151 \times 10^{-6} \text{ m}^3$$

$\Rightarrow$
$$\sigma_{max} = \frac{M_{max}}{Z_{xx}} = \frac{126.07}{7.151 \times 10^{-6}}$$

$$\sigma_{max} = 17.63 \text{ MPa (which is safe)}$$

Maximum compressive stress in connecting rod

$$= \frac{F_c}{A} = \frac{13,343.2}{11 \times (8 \times 10^{-3})^2}$$

$$= \frac{F_c}{A} + \sigma_{max}$$

$$= \frac{13,343.2}{11(8 \times 10^{-3})^2} + 17.63 \times 10^6$$

$$= (18.95 + 17.63) \times 10^6 \text{ N/m}^2 = 36.58 \text{ MPa} \qquad \text{(which again is safe)}$$

Small End Dimensions

$F_p = l_p d_p P_b$

l_p = Length of piston pin

d_p = Diameter of piston pin

P_b = Permissible bearing pressure

Taking $\dfrac{l_c}{d_c} = 1.50$ and $P_b = 10$ MPa

$\Rightarrow$ $21{,}276.64 = 1.5 d_p \times d_p \times 10$ or $d_p \approx 38$ mm

$\Rightarrow$ $l_p = 1.5 d_p = 1.5 \times 38 = 57 \approx 58$ mm

Big End Diameter

$$F_p = l_c d_c P_b$$

Taking $\dfrac{l_c}{d_c} = 1.25$ and $P_b = 7.5$ MPa

$\Rightarrow$ $21{,}276.64 = 1.25 d_c = 47.6 \approx 48$ mm

$$l_c = 1.25 \times 18 = 60 \text{ mm}$$

Big End Bolts

$$F_i = 2 \times \frac{\pi}{4} d_b^2 \times \sigma_t.$$

Taking $\sigma_t = k$ N/mm^2

or $7{,}934.5 = 2 \times \dfrac{\pi}{4} \times d_b^2 \times 12 \Rightarrow d_b = 20.52$ mm

So we use M_{22} size bolts.

or $d_b = 22$ mm

Big End Cap

$M_{max} = \dfrac{F_i l_0}{6} \Rightarrow l_0 = d_c + 2 \times \text{Thickness of linear} + d_b + \text{Clearance of 1.5 mm (say)}$

$$= d_c + 2 \times (0.05 d_c + 1) + d_b + 1.5$$
$$= 48 + 2(0.05 \times 48 + 1) + 22 + 1.5$$
$$= 78.3 \text{ mm}$$

$$M_{max} = \frac{F_i l_0}{6} = \frac{7{,}934.5 \times 78.3 \times 10^{-3}}{6}$$

$$M_{max} = 103.54 \text{ N-m}$$

$$M_{max} = \sigma_b \times \frac{l_c l_r^2}{6}$$

Taking $\sigma_b = 120$ N/mm^2

$$\text{Cap thickness}\,(h) = \sqrt{\frac{6 M_{max}}{\sigma_b l_c}}$$

$$h = \sqrt{\frac{6 \times 103.54}{120 \times 60 \times 10^{-3}}}$$

$$h = 9.28$$

say $h = 10$ mm

2. The cylinder of a slow speed steam engine is 250 mm diameter and the steam pressure is 1 N/mm². The piston rod length is 1,000 mm and the connecting rod is 1.2 m long. The engine stroke is 550 mm. Determine the dimensions of the cross-section of the connecting rod assuming the depth to be twice as thickness and a suitable diameter for the piston rod.   (UPTU 2011)

SOLUTION: $N = 200$ rpm, $P = 1$ N/mm², $d = 250$ mm, $l = 1.2$ m, stroke = 550 mm

$H = 2t$, $B = t$

$A = 2(2t \times t) + t \times t = 5t^2$

$$I_{xx} = \frac{1}{12}\left[2t(t)^3 - t^4\right] = \frac{t^4}{12}$$

$$I_{yy} = 2 \times \frac{1}{12} \times t \times (2t)^3 + \frac{1}{12} \times t \times t^3 = \frac{t^4}{6} + \frac{t^4}{12} = \frac{t^4}{4}$$

$$\frac{I_{xx}}{I_{yy}} = \frac{t^4}{12} \times \frac{12}{17t^4} = \frac{1}{17}$$

$$F_c = F_L = \frac{\pi D^4}{4} \times P = \pi \times \left(\frac{250}{1,000}\right)^2 \times 1 = 0.049 \text{ N}$$

$$K_{xx} = \sqrt{\frac{I_{xx}}{A}} = \sqrt{\frac{t^4}{12} \times \frac{1}{5t^2}} = \frac{t^2}{60} = \frac{t}{7.75}$$

Length of crank $(r) = \dfrac{\text{Stroke of piston}}{2} = \dfrac{550}{2} = 275$ mm

Length of connecting rod $(l) = 1.2$ m $= 1,200$ mm

According to Rankine formula

$$0.049 = \frac{320 \times 5t^2}{1 + \dfrac{1}{7,500}\left(\dfrac{1,200 \times 7.75}{t}\right)^2} = \frac{320 \times 5 \times t^2}{1 + \dfrac{11,532}{t^2}} = \frac{1,600t^4}{\left(t^2 + 11,532\right)}$$

$\Rightarrow \qquad 0.049t^2 + 565.068 = 1,600t^4$

$t^2 = 47.6$

$\therefore t = 7$ mm, $H = 14$ mm, $B = 7$ mm

6.7 DESIGN OF CENTER CRANKSHAFT

Center crankshafts are designed by considering two crank positions, i.e. when the crank is at dead center (or when the crankshaft is subjected to maximum bending moment) and when the crank is at an angle at which twisting moment is maximum.

Case A: When crank is at dead center

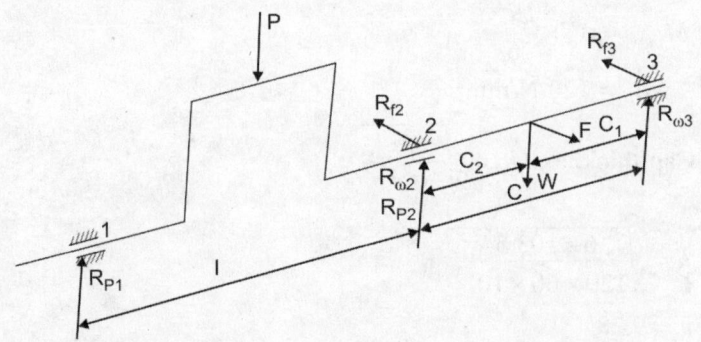

Fig. 6.16 Crankshaft subjected to forces at TDC position

At this position of the crank, the maximum gas pressure on the piston will transmit maximum force on the crank pin in the plane of the crank causing only bending of the shaft.

Considering from figure, crank being supported at three main bearings 1, 2 and 3. The crankshaft at the *TDC* position is subjected to gas force *P*, weight of flywheel *W* and the sum of the tight side and slack side tensions *F*.

Assuming the gas force is concentrated at the midpoint of the crank pin, the radiation of force *P* on the bearings 1 and 2 are: $R_{P_1} = R_{P_2} = P/2$

The reactions at bearings 2 and 3 due to flywheel weight are $R_{\omega_2} = \dfrac{W_{C_1}}{2}$ and $R_{\omega_3} = \dfrac{W_{C_2}}{C}$

Horizontal reactions at bearings 2 and 3 due to bolt tension are: $R_{f_2} = \dfrac{FC_1}{C}$ and $R_{f_3} = \dfrac{FC_2}{C}$

The resultant reaction forces at bearings 2 and 3 are

$$R_2 = \sqrt{R_{f_2}^2 + \left(R_{P_2} + R_{\omega_2}\right)^2}$$

$$R_3 = \sqrt{R_{f_3}^2 + R_{\omega_3}^2}$$

Design of Crank Pin

The crank pin at *TDC* is subjected to bending moment, bearing pressure and shear force.

Bending stress $(\sigma_b) = M/Z$

where maximum bending moment on the crank pin

$$M = R_{P_1} \times l/2$$

and

$$Z = \frac{\pi}{32}\left(d_{pin}\right)^3$$

Bearing pressure $(P_{br}) = \dfrac{\text{Load}}{\text{Projected area}} = \dfrac{P}{l_{pin} \times d_{pin}}$

For modern automobile engines, the ratio of the crank pin length to diameter is generally between 0.8 and 1.1 considering double shear failure of pin.

$$\tau = \frac{P}{\left(2 \times \dfrac{\pi}{4} d_{pin}^2\right)}$$

Table 6.2 Allowable bearing pressure

Class of work	Main bearing pressure (P_{br}) (N/mm²)	Crank pin pressure (P_{br}) (N/mm²)
Automobile engines	10–14.0	2.5–3.0
Diesel engines	5–7.0	7–9.0
Marine diesel engine	2.5–3.0	7–10.0
Rail road, locomotive	1.2–1.4	10–12.0
Shear and punches	20–27	34–54.0

Design of Crank Web

The crank web at *TDC* position is subjected to bending moment and direct compressive stresses.

Empirical relations for the dimensions of the crank web:

Width $(b) = (1.1 \text{ to } 1.2)d_{pin}$

Thickness $(t) = (0.6 \text{ to } 0.75)d_{pin}$

Bending stress $(\sigma_{b_1}) = \dfrac{M_1}{Z_1} = \dfrac{\dfrac{P}{2}\left[\dfrac{l}{2} - \dfrac{l_{pin}}{2} - \dfrac{t}{2}\right]}{\dfrac{1}{6}bt^2}$

Direct compressive stress due to gas force

$$\sigma_c = P/bt$$

So total stress on the crank web

$$= \text{Bending stress} + \text{Direct stress}$$

$$= (\sigma_{b_1} + \sigma_c)$$

This total stress should be less than the permissible bending stress.

Design of Shaft Under the Flywheel

The shaft under the flywheel is subjected to maximum bending moment at the flywheel location, i.e.

$$M_{shaft} = R_3 C_1 = \frac{\pi}{32} d_{shaft}^3 \sigma_b$$

where d_{shaft} = Shaft diameter

σ_b = Allowable strength

Case B: When crank is at an angle of maximum twisting moment.

The twisting moment on the crankshat will be maximum when the tangential force on the crank is maximum.

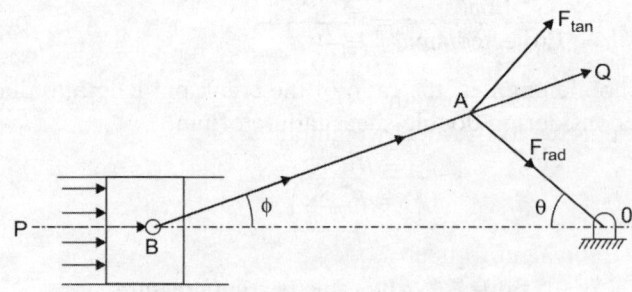

Fig. 6.17 Crank at maximum torque position

The maximum value of tangential force lies when the crank is at an angle of 25° to 30° from the dead center for constant volume combustion engines (i.e. petrol engines) and 30° to 40° for constant pressure combustion engines (i.e. diesel engines).

At this crank angle, the gas pressure is not maximum. Let the gas force acting at that instant be P'.

Then $Q = \dfrac{P'}{\cos\phi}$

where ϕ = Obliquity angle = $\sin^{-1}\left[\dfrac{\sin\theta}{n}\right]$

$$n = \frac{l_n}{r}$$

Tangential force $(F_{tan}) = Q \sin(\theta + \phi)$
and the radial force along the crank

$$F_{rad} = Q \cos(\theta + \phi)$$

Tangential force F_{tan} will have two reactions R_{tan1} and R_{tan2} at bearings 1 and 2 respectively.

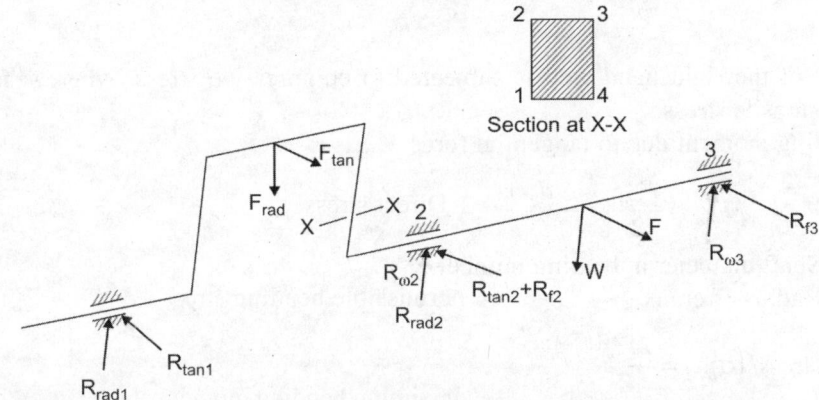

Fig. 6.18 Crank inclined at angle θ

Similarly the radial force F_{rad} will have two reactions R_{rad1} and R_{rad2}. Reactions at bearings 2 and 3 due to flywheel weight and belt tension remain the same.

The free body diagram of crankshaft with various forces (shown above). Crankshaft in this position is subjected to combined bending moment and torque.

If P' is the intensity of pressure on the piston at this instant, the bending moment due to the radial component is $(M) = R_{rad1} \times l/2$
and torque $(T) = R_{tan1} \times r$

So equivalent torque $\left(T_{eq}\right) = \sqrt{M^2 + T^2} = \frac{\pi}{16} d_{pin}^3 \times \tau$

where 'τ' is the allowable shear strength.

The bending moment due to flywheel weight and belt tension is

$$M_{shaft} = R_3 C_1$$

and torque

$$(T) = F_{tan} \times r$$

Equivalent torque $= \sqrt{M_{shaft}^2 + T^2} \le \frac{\pi}{16} d_{shaft}^3 \times \tau$

Now we need to find shaft diameter at the junction of right hand crank web,

Bending moment

$$M = R_1 \left[\frac{l}{2} + \frac{l_{pin}}{2} + \frac{t}{2} \right] - Q \left[\frac{l_{pin}}{2} + \frac{t}{2} \right]$$

where R_1 is the resultant reaction at bearing 1. Torque $(T) = F_{tan} \times r$

Equivalent torque $= \sqrt{M^2 + T^2} \le \frac{\pi}{16} d_2^3 \tau.$

where d_2 is the diameter of the shaft at the junction of the right hand web.

The right hand crank web is subjected to severe load condition. Various stresses induced are:

The bending moment due to radial component is

$$M_{rad} = R_{rad} = \left[\frac{l}{2} - \frac{l_{pin}}{2} - \frac{t}{2} \right]$$

Bending stress $\left(\sigma_{b_1} \right) = \dfrac{M_{rad}}{Z}$

where $\qquad Z = \dfrac{bt^2}{6}$

Face 1–2 of the right hand web is subjected to compressive stress, whereas face 3–4 is subjected to tensile stress.

The bending moment due to tangential force F_{tan} is

$$M_{tan} = \frac{F_{tan}}{2} \left[r - \frac{d_2}{2} \right]$$

where d_2 = Shaft diameter at bearing number 2
$\qquad r$ = Radius of crank

Bending stress $\left(\sigma_{b_2} \right) = \dfrac{M_{tan}}{Z}$

where $\qquad Z = \dfrac{tb^2}{6}$

Here σ_{b_2} is the compressive stress on face 1–4 and tensile stress on face 2–3.

Radial force F_{rad} also produces direct compressive stress, i.e. $\sigma_c = \dfrac{F_{rad}}{2bt}$

Thus maximum compressive stress

$$\sigma_{c'} = \sigma_{b_1} + \sigma_{b_2} + \sigma_c \le \sigma_{allowable}$$

Crank web is subjected to twisting moment due to tangential force

$$T = \frac{F_{tan}}{2} \left[\frac{l_{pin}}{2} + \frac{t}{2} \right]$$

Torsional shear stress $(\tau) = \dfrac{T}{Z_{polar}}$

Z_{polar} = Polar section modulus = $\dfrac{bt^2}{4.5}$

Since web is subjected to biaxial stresses, thus the maximum principal stress should be within the allowable limit.

$$\sigma_1 = 0.5 \left[\sigma_c' + \sqrt{\sigma_c'^2 + 4\tau^2} \right] \le \sigma_{allowable}$$

The left hand crank is less severely stressed than the right hand crank web. Thus dimensions of the left hand web may be taken the same as those of the right hand web.

WORKED EXAMPLES (PREVIOUS YEAR UNIVERSITY QUESTIONS)

1. For a single cylinder 4-stroke petrol engine, design the crank pin and crank webs of a center crankshaft for the engine with parameters given below:

Cylinder diameter = 25 cm

Stroke = 40 cm

Maximum explosion pressure = 2.4 MPa at *TDC*

Gas pressure at maximum torque position when crank angle is 40° = 0.9 MPa

Weight of flywheel = 16 kN

Total tension in belts on tight and slack side = 3.5 kN

Length of connecting rod = 95 cm

Speed = 1,000 rpm

(UPTU 2002)

SOLUTION: We shall design the crankshaft for two positions of crank, i.e. firstly when the crank is at the dead center and secondly when the crank is at an angle of maximum twisting moment.

Design of crank pin when crank is at the dead center:

$$\text{Piston gas load} = \frac{\pi}{4}d^2 \times P = \frac{\pi \times (250)^2}{4} \times 2.4$$

$$= 117{,}750 \text{ N} = 117.75 \text{ kN}$$

Assuming that the distance (*b*) between bearings 1 and 2 is equal to twice the piston diameter (*d*).

$$b = 2d = 2 \times 400 = 800 \text{ mm}$$

$$b_1 = b_2 = \frac{b}{2} = 400 \text{ mm}$$

Due to piston gas load, there will be two horizontal reactions H_1 and H_2 at bearings 1 and 2 respectively, such that

$$H_1 = \frac{F_p \times b_1}{b} = \frac{117.75 \times 400}{800} = 58.88 \text{ kN}$$

$$H_2 = \frac{F_p \times b_2}{b} = \frac{117.75 \times 400}{800} = 58.88 \text{ kN}$$

Assume that length of main bearings to be equal, i.e. $C_1 = C_2 = C/2$. We know that due to the weight of the flywheel acting downwards, there will be two vertical reactions V_2 and V_3 at bearings 2 and 3 respectively, such that

$$V_2 = \frac{W \times C_1}{C} = \frac{W \times C/2}{2} = \frac{W}{2} = \frac{16}{2} = 8 \text{ kN}$$

$$V_3 = \frac{W \times C_2}{C} = \frac{W \times C/2}{2} = \frac{W}{2} = \frac{16}{2} = 8 \text{ kN}$$

Due to the resultant belt tension $(T_1 + T_2)$ acting horizontally, there will be two horizontal reactions H_2 and H_3 respectively, such that

$$H_2 = \frac{(T_1 + T_2)C_1}{C} = \frac{T_1 + T_2}{2} = \frac{3.5}{2} = 1.75 \text{ kN}$$

$$H_3 = \frac{(T_1 + T_2)C_2}{C} = \frac{T_1 + T_2}{2} = \frac{3.5}{2} = 1.75 \text{ kN}$$

Crank pin and web is designed as discussed below:

a. Design of crank pin

Let d_c = Diameter of crank pin in mm

l_c = Length of crank pin in mm

σ_b = Allowable bending stress for crank pin

It may be assumed as 75 MPa for selected material. We know that the bending moment at the center of crank pin.

$$M_c = H_1 \times b_2 = 58.88 \times 400 = 23{,}552 \text{ kN-mm} \qquad \dots(1)$$

We also know that

$$M_c = \frac{\pi}{32}(d_c)^3 \sigma_b = \frac{\pi}{32}(d_c)^3 \times 75 = 7.364(d_c)^3 \text{ N-mm}$$

$$M_c = 7.364 \times 10^{-3}(d_c)^3 \text{ kN-mm} \qquad \dots(2)$$

Equating (1) and (2), we get

$$23{,}552 = 7.364 \times 10^{-3}(d_c)^3$$

$$d_c = 147.7 \text{ mm}$$

$$d_c = 148 \text{ mm}$$

We know that length of crank pin

$$l_c = \frac{F_p}{d_c \times P_b} = \frac{117.75 \times 10^3}{148 \times 8} \qquad \text{(taking } P_b = 8 \text{ N/m}^2\text{)}$$

$$l_c = 99.56, \text{ say } 100 \text{ mm}$$

b. Design of left hand crank web thickness of crank web

$$t = 0.65 d_c + 6.35$$

$$t = 0.65 \times 148 + 6.35 = 102.55, \text{ say } 103 \text{ mm}$$

Width of crank web

$$W = 1.125 d_c + 12.7$$

$$W = 1.125 \times 148 + 12.7 = 179.2 = 180 \text{ mm.}$$

We know that maximum bending moment on crank web

$$M = H_1\left(b_2 - \frac{l_c}{2} - \frac{t}{2}\right) = 58.88\left(400 - \frac{100}{2} - \frac{10}{2}\right)$$

$$M = 17{,}576 \text{ kN-mm}$$

Section modulus $(Z) = \dfrac{1}{6} wt^2 = \dfrac{1}{6} \times 180 \times (103)^2$

$$Z = 318.3 \times 10^3 \text{ mm}^3$$

Bending stress $(\sigma_b) = \dfrac{M}{Z} = \dfrac{17{,}576 \times 10^3}{318.3 \times 10^3} = 55.2 \text{ N/mm}^2.$

Direct compressive stress on crank web

$$\sigma_c = \frac{H_1}{Wt} = \frac{58.88 \times 10^3}{180 \times 103} = 3.2 \text{ N/mm}^2$$

Total stress on the crank web = $\sigma_b + \sigma_c = 55.22 + 3.2$

$$\sigma_{total} = 58.4 \text{ N/mm}^2 \text{ or } 58.4 \text{ MPa}$$

Since total stress on crank web is less than allowable bending stress of 75 MPa, therefore, design of left hand crank web is safe.

c. Design of right hand crank web

From balancing point of view, dimensions of right hand crank web (thickness and width) are made equal to the dimensions of the left hand crank web (2). Design of crankshaft when crank is at an angle of maximum twisting moment.

$$\text{Piston gas load} = \frac{\pi}{4} \times D^2 \times S'$$

$$F_P = \frac{\pi}{4} \times (400)^2 \times 0.9 = 113.1 \text{ kN}$$

In order to find the thrust in connecting rod (F_Q) we should first find out the angle of inclination of connecting rod with line of stroke.

$$\text{We know that } \sin\phi = \frac{\sin\theta}{l/r} = \sin 40°$$

$$\frac{l}{r} = \frac{950}{200} = 4.75, \text{ where } r = \frac{S}{2}$$

$$\therefore \quad \sin\phi = \frac{\sin(40)}{4.5} = 0.1428$$

$$\phi = 8.20°$$

$$\text{Thrust in connecting rod } \left(F_Q\right) = \frac{F_P}{\cos\phi} = \frac{113.1}{\cos(8.2)}$$

$$F_Q = 114.3 \text{ kN}$$

Tangential force acting on crankshaft

$$F_T = F_Q \sin(\theta + \phi) = 114.3 \sin(40 + 8.2)$$

$$F_T = 85 \text{ kN}$$

Radial force $(F_R) = F_Q \cos(\theta + \phi) = 114.3 \cos(48.2)$

$$F_R = 76.18 \text{ kN}$$

Due to tangential force (F_T) there will be two reactions at the bearings 1 and 2, such that

$$H_{T_1} = \frac{F_T \times b_1}{b} = \frac{85 \times 400}{800} = 41.5 \text{ kN}$$

$$H_{T_2} = \frac{F_T \times b_2}{b} = \frac{85 \times 400}{800} = 42.5 \text{ kN}$$

Due to radial force (F_R), there will be two reactions at the bearings 1 and 2, such that

$$H_{R_1} = \frac{F_R \times b_1}{b} = \frac{76.18 \times 400}{800} = 38.1 \text{ kN}$$

$$H_{R_2} = \frac{F_R \times b_2}{b} = 38.1 \text{ kN}$$

Now design of crank pin is done as below:

a. Design of crank pin

Let d_c = Diameter of crank pin in mm

Bending moment at center of crank pin

$$M_c = H_{R_1} \times b_2 = 38.1 \times 400 = 15,236 \text{ kN-mm}$$

Twisting moment of crank pin

$$T_c = H_{T_1} \times r = 42.5 \times 200 = 8{,}500 \text{ kN-mm}$$

Equivalent twisting moment

$$T_e = \sqrt{M_c^2 + T_c^2} = \sqrt{(15{,}236)^2 + (8{,}500)^2} = 17{,}446.6 \text{ kN-mm}$$

We know that equivalent twisting moment (T_e)

$$17.446 \times 10^6 = \frac{\pi}{16}(d_c)^3 \times \tau$$

On solving we get

$$d_c = 136.4 = 137 \text{ mm}$$

Since this value of crank pin ($d_c = 137$ mm) is less than the already calculated value of d_c (i.e. $d_c = 148$ mm), therefore, we shall take 'd_c' of crank pin = 148 mm.

KEY TERMS

- Introduction
- Cylinder
- Design of cylinder
- Piston
- Design of piston
- Connecting rod
- Crank shaft
- Design of center crankshaft

SUMMARY

An IC engine has the following major components like cylinder block, cylinder lines, cylinder head, piston, piston rings, piston pin, connecting rod, crankshaft, etc. Classification is on the basis of number of strokes, type of operating cycle and arrangement of cylinders.

Design criteria as well as design parameters for different parts of IC engine, specifically cylinder, piston, connecting rod and crankshaft are discussed in this chapter. Design includes criteria for different components of these IC engine parts like cylinder liner, piston pin, piston ring, connecting rod with small and big end and crank web and crank pin respectively.

Various stresses acting on IC engine components, viz. cylinder, piston, connecting rod and crankshaft are taken into consideration at different positions of operations like TDC (top dead center) and BDC (bottom dead center).

REVIEW QUESTIONS

1. Why piston clearances are necessary? What is its usual value?
2. Explain the effect of piston crown thickness and diameter or heat flow in brief and lubrication of piston rings.
3. Explain stresses induced in connecting rods and crankshaft.
4. Briefly explain design considerations of piston.
5. Determine selection criteria of type of IC engine.
6. Explain valve gear mechanism of IC engine.
7. Explain various stresses and failure criteria induced in the connecting rod of an IC engine.

8. Following data refers to a single cylinder 4 stroke petrol engine.
   Cylinder diameter = 60 mm
   Stroke = 100 mm
   Maximum explosion pressure = 2.1 MPa at 15° angle from TDC gas pressure at maximum
   Torque position when crank angle is 40° = 0.8 MPa
   Weight of flywheel = 5 kN
   Length of connecting rod = 225 m
   Speed = 800 rpm
   Design a connecting rod having I-section or the crank pin and crank webs of the center
   crank shaft for the engine.

9. Design a connecting rod for 4 stroke petrol engine, with the following data:
   Piston diameter = 0.10 m
   Stroke length = 0.15 m
   Length of connecting rod (center to center) = 0.30 m
   Weight of reciprocating parts = 20 N
   Speed is 1,500 rpm with possible overspeed of 2,500 rpm, compression ratio = 4:1,
   maximum explosion pressure = 2.5 MPa

10. Design a piston for a single acting four stroke engine for the following specifications.
    Cylinder bore = 0.30 m, stroke length = 0.375 m
    Maximum gas pressure = 8 MPa
    Brake mean effective pressure = 1.15 MPa
    Fuel consumption = 0.22 kg/kW/hour
    Speed = 500 rpm. Assume suitable data.

Index